Spray
Polyurethane
Foam in
External
Envelopes
of Buildings

Spray Polyurethane Foam in External Envelopes of Buildings

MARK T. BOMBERG
JOSEPH W. LSTIBUREK

CRC Press
Taylor & Francis Group
Boca Raton London New York

CRC Press is an imprint of the
Taylor & Francis Group, an **informa** business

CRC Press
Taylor & Francis Group
6000 Broken Sound Parkway NW, Suite 300
Boca Raton, FL 33487-2742

First issued in paperback 2019

ISBN-13: 978-1-56676-707-1 (hbk)
ISBN-13: 978-0-367-40028-6 (pbk)
Library of Congress Card Number 98-86273

Library of Congress Cataloging-in-Publication Data

Main entry under title:
Spray Polyurethane Foam in External Envelopes of Buildings

Full Catalog record is available from the Library of Congress

Visit the Taylor & Francis Web site at
http://www.taylorandfrancis.com

and the CRC Press Web site at
http://www.crcpress.com

CONTENTS

THE CONSTRUCTION INDUSTRY undergoes slow but continuous change. The current welcome, and long overdue, trend is to quantify performance of the whole building system, rather than that of individual products and components. Objective based codes and performance-oriented standards are being developed. Cost-benefit analysis replaces prescriptive recommendations.

This book started as a dialog between two of us, each having different backgrounds in the field of building science—a dialog on applicability of the concept of performance. One soon realizes that the only way to define performance of materials is to define the onset of failure. Materials, however, may contribute to the system performance in many different ways and failure of a building component is usually affected by a multitude of simultaneously acting factors and effects.

At the same time, one of us was asked by the Spray Polyurethane Foam industry to define the aspects of material performance that slow the growth of this industry in wall system applications. This focused our dialog into a more rigorous analysis.

Why have the authors selected sprayed polyurethane foam (SPF) for this research review? For one, SPF as a system has intrigued the authors over their professional careers. A continuous SPF layer performs functions of an environmental barrier (controlling heat, air, and moisture flows through the building envelope), while simultaneously transfering loads (snow, wind) and structural movements (material shrinkage or expansion).

To review field performance of SPF, one must include many different aspects of material and system performance. The authors believe their approach to SPF is a model equally valid for other material systems.

The moisture performance of SPF is rather complex. For instance, the layer of SPF may either be a "retarder," reducing the rate of vapor flow, or a "breather," allowing vapor to pass through the SPF layer. The SPF layer may be fabricated with different resistance to vapor flow, depending on environmental conditions during the application, the manner in which this material is applied, and the nature of the substrate. Presence or absence of temperature gradients will also affect the amount of moisture accumulating with the SPF layer.

The 1995 National Building Code (the model code for Canada) states that an assembly can be designed either to prevent condensation from occurring at all or keep it from accumulating and causing deterioration. Moisture design of structures containing SPF should be based on the second option, namely on the "flow through" approach. In this approach, the amount of moisture which may be accumulated during one season is assessed with a view to two requirements:

- Can this moisture affect durability of materials contained in the building envelope?
- Can this moisture dry out without any increase from year to year?

Moisture engineering leads to both "theoretical" and "practical" aspects of building science. The theoretical aspects include principles of design for durability and long-term thermal and moisture performance of materials. The "practical" aspects must recognize differences in moisture performance of various SPF products. The result is a monograph consisting of two parts.

Part 1 of this monograph analyzes SPF performance as the material (product). Being field-fabricated, installation of SPF products must include a quality assurance (QA) program, which in turn should be based on an understanding of factors affecting the performance of SPF. Laboratory evaluation of foams and their coverings, quality management issues, and quantification of the technical support provided to the SPF contractor are also reviewed in this book.

Part 2 presents a systems approach to construction. Starting with principles of environmental control of buildings, different aspects of design and performance of roofing and wall systems are reviewed. Details and design recommendations based on designs obtained from leading architects, designs published in architectural practice magazines as well as case studies where the SPF is used for rehabilitation of the building envelope are included.

It is our intention that this monograph leads from principles, through the architectural design, material selection, and construction, to the commissioning and acceptance of construction.

THE AUTHORS EXPRESS deep gratitude to the National Research Council of Canada and Canadian Plastics Industry Association for permission to reprint many data from the joint NRC/CPIA research project on CFC replacement in SPF products. Special thanks are due to Dr. Kumar Kumaran, head of the Thermal Insulation Laboratory, NRCC and to Art Pazia, Chem-Thane Inc., for permission to reprint or modify many of the architectural details from the Chem-Thane Manual on Spray Foam. Thanks are forwarded to Sandra Marshall, CMHC, for permission to reprint/modify selected architectural details from "Best practice guide for wood frame walls." We are grateful to Anthony Woods, Flexible Products, for providing us with almost thirty case studies where sealing foams were used and many discussions. Gratitude is expressed to Joe Sartor, Contec; Mason Knowles, former technical director of SPFD/SPI; David Gluck, formerly with Jim Walter Research; to Jim Andersen, Darryl Bennet and George Sievert, Foam Enterprise Inc.; to Mario Charlebois and Dave Lall, Demilec Inc., who together with many other people within the SPF industry had significant input to this manuscript.

PART ONE

Performance Evaluation of System Components

CHAPTER 1

Definition of Spray
Polyurethane Foams

1. INTRODUCTION

THE TERM "SPRAY polyurethane foam" (SPF) represents a general class of foams (cellular plastics) fabricated as a product of a catalyzed chemical reaction of polyisocyanate (component A) and polyhydroxyl compounds (component B). The latter component also contains other compounds such as stabilizers, nucleating agents, and blowing agents. The SPF products are fabricated in situ. In most cases, these products are applied directly on the surface to be insulated or sealed for air transfer through the construction. In addition to chemically cured, two-component foams, the SPF category also includes one-component, moisture-cured foams, often called polymeric sealants.

2. SPRAY POLYURETHANE FOAM

SPF as a category comprises either closed-cell or open-cell structures with densities ranging from 8 to 65 kg/m³ (0.5 to 4 lb/ft³). While the foam density cannot be used as the criterion of foam performance, the core density is extensively used in quality assurance programs.

In a typical two-component foam fabrication, two transfer pumps deliver components A and B to the proportioning unit. Then, a heated dual-hose carries each of the components to the mixing gun. As a rule, A and B components are delivered to the spray gun in a one-to-one ratio. Compressed air is also delivered to the mixing chamber, though via a separate, unheated hose. Depending on the reactivity of the polymeric mixture and the delivery mechanism, SPF may be recognized as:

- poured foam, when a liquid stream reaches the substrate
- spray foam, when a plethora of small liquid droplets reach the substrate
- froth foam, when the liquid droplets reaching the substrate contain already nucleated miniature gas bubbles

A mechanism used to discharge a one-component foam from a pressurized container is different. A propellant gas extrudes a bead of the foam through a plastic (removable) nozzle placed on the container with the foam. In principle, a two-component foam can also be applied in the form of a narrow strip, making the difference between container and gun applied foams a mere "technicality."

Since SPF can perform varied functions such as bonding, thermal insulation, and air or moisture control in buildings, it becomes a preferred choice for many applications. SPF products are used for insulating roofs, storage tanks, ducts, and pipes (district heating, underground pipes). They are also used in sewer renewals (where the bonding strength of specially developed foams is well utilized). A case of house masonry rehabilitation with corroded masonry ties is discussed by Woods (1987). Instead of demolishing the outer leaf of the brick wall to install new wall-ties, the rigid SPF was injected to bond the brick veneer to the inner wall. Similarly, SPF products can also be used to consolidate friable strata in mining operations or as a measure to prevent soil erosion. Specially designed, poured SPFs have also been used for repair (leveling) of industrial floors where a slab on ground construction has experienced uneven settlement.

In construction, SPF products are mainly used to control heat transfer through the building envelope. However, the replacement of traditional blowing agents, namely chlorofluorocarbons (CFCs) with partially halogenated hydrochlorofluorocarbons (HCFCs), which have thermal insulating performance inherently lower than that of CFCs, has raised the question of the long-term thermal resistance (LTTR) of such foams.

Research (Bomberg and Kumaran, 1989) showed that HCFC-141b blown foam can be optimized to have long-term thermal performance almost identical to a CFC blown foam. The same research has also shown that some commercially available SPF systems were not optimized with regard to long-term thermal performance. One poorly optimized product is shown in Figure 1 against the same SPF product that was previously manufactured with CFC-11.

Figure 1 shows that initially, foams exposed in the laboratory had a slightly different rate of aging than those in the service conditions. Yet, after a few months, these differences disappear. In effect, aging in the laboratory may be used to represent aging under field conditions. This important observation was also confirmed in other studies (Christian et al., 1991).

So, the significant difference between products P1 and P2 shown in Figure 1 cannot be ascribed to differences in testing or to effects of exposure. This difference is clearly related to the product. Since the polymeric composition was identical, this difference may be associated with the blowing agent used.

Another important question is asked by Willingham (1991): whether or not a "product that is essentially prepared and applied on site, will conform to the same specifications and standards each time it is installed."

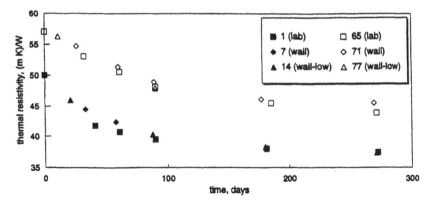

FIGURE 1. Thermal resistivity (inverse of thermal conductivity) of the SPF product manufactured with CFC-11 (open symbols) and HCFC-141b (filled symbols). Samples stored in the laboratory are compared with those exposed in the walls of the experimental station. Reprinted with permission of the National Research Council, Canada.

3. CONSISTENCY OF THE SPF FABRICATION PROCESS

Table 1 shows density measured on 610 mm 610 mm (2' × 2') specimens cut from 1.2 m × 1.2 m (4' × 4') samples. These samples were sprayed on the site of a large renovation project, where SPF was used as the wall insulation. To establish consistency of the field application, each of the samples was sprayed during a different day and sent for testing to a nearby laboratory.

A sensitivity study was also performed on the site to assess the effect of alterations in the installation conditions. Such effects as changing the mixing chamber temperature, gun pressure, and distance from the substrate were examined. Six samples were collected and tested in the same laboratory. The SPF density varied between 36.5 kg/m³ (2.28 lb/ft³) and 39.1 kg/m³ (2.44 lb/ft³); the average was 38.0 kg/m³ (2.35 lb/ft³). This average density was practically identical to the one obtained for random samples (Table 1).

While the average density remained invariable, the scatter of results shown in Table 1 was larger than that measured when systematically changing the installation variables. Subsequently, a detailed study of variability in the SPF fabrication process was undertaken.

Two characteristics are used to define variations in the fabrication process: repeatability and reproducibility. Repeatability is variation in properties produced when different batches were made by one operator using the same SPF product and spray equipment. Reproducibility is the variation between different operators and spray equipment. The spray foam equipment was characterized in Table 2 and conditions of foam installation at each plant are listed in Table 3.

Using these two benchmarks, each of the three manufacturers (system houses) participating in the joint project adjusted either the equipment or the SPF foaming conditions until the core density was within 10% of that originally planned. Though the same delivery system and proportioning units were used at two of the three manufacturing locations, Table 2 shows that the conditions in the mixing chamber were slightly different.

Six different product batches were then manufactured over a period of two months (three at each plant) to examine repeatability and reproducibility of the SPF. Substitution of the component A (marked as batch 12A, see Bomberg and Kumaran, 1989) was also examined. Samples from each product batch were taken for measuring physical and thermal characteristics of the material.

Table 4 shows basic characteristics of the foaming process as measured in two locations. Some of the results obtained from batches 13 and 14 are also shown in Table 4. Results listed in Table 4 indicate that density determined on thin layers may be used to detect denser layers (surface skins), which affects mechanical performance of the foam. The density, however, is not correlated with the thermal performance of the foam.

Density alone is not sufficient to quantify how well the field fabrication process can be repeated. Other aspects of foam performance must also be quantified. For instance, an initial thermal resistivity of the SPF (thermal resistance of a unit thickness) may be used for this purpose. Table 5 uses this approach and shows the thermal resistivity of SPF specimens cut from batches manufactured at two different locations.

Consistency of these results can be compared with variability in thermal properties of a laminated polyisocyanurate (PIR) product examined by Sherman (1980). Standard deviations of 1.4% and 2.5%, shown in Table 4, may be compared with the corresponding standard deviations in the thermal measurements of PIR, namely: 2.8%, 4.4%, 4.5%, and 6.2% (Sherman, 1980).

In effect, thermal performance of the SPF manufactured in the collaborative industry/NRCC project had better repeatability than that reported for a PIR lamination process. A high repeatability of initial SPF thermal performance was also shown by Bomberg (1993) for five batches of two commercial SPF products manufactured by three qualified contractors.

The answer to the question posed by Willingham (1991) is positive. The spray foam that is fabricated on the construction site "will conform to the same specifications and standards each time it is installed" if, during foam fabrication process, the installer controls all critical variables. High consistency

Table 1. The overall density in kg/m³ (lb/ft³)
determined on samples from different SPF batches.

1	39.1 (2.44)	5	38.5 (2.39)
2	32.2 (2.01)	6	31.7 (1.98)
3	41.3 (2.58)	7	38.7 (2.42)
4	38.6 (2.41)	8	39.9 (2.49)

The average of these eight measurements gave 37.5 kg/m³ (2.34 lb/ft³).

Table 2. Characteristics of the spray equipment in the
comparative study (Bomberg and Kumaran, 1989).

Description	Location 1	Location 2
Delivery system		
Proportioning unit	Model H-II	Model H-II
Gun type	Model D	Model D
Mixing chamber conditions		
Static primary termperature	48°C (120°F)	52°C (125°F)
Range of static pressures	56–70 atm (800–1200 psi)	56–70 atm (800–1200 psi)
Hose temperature	35°C (95°F)	38° C (100°F)
Hose length	8 m (25 ft)	12 m (35 ft)

Table 3. Characteristics of foaming process in the discussed comparative study.

1. Cream and tack-free times
2. Maximum temperature reached in the foam
3. Time when the maximum temperature occurred
4. Density measured on a 25-mm thick (1 inch) layer cut from the foam core (ASTM D 1622)
5. Density of 10–14 mm thin layers (1/2 inch) cut from the foam core
6. Density of such a layer cut from a layer including two spray passes
7. Distance from the installer foot to the substrate to ensure that all batches were produced with a similar distance between the spray gun and substrate

Table 4. Characteristics of the foaming process,
measured in both manufacturing locations.

Batch/Location	Cream Time, s	Tack-Free Time, s	Maximum Temp., °C	Time (min) at Max. Temp.	Overall Density* kg/m³	Density Core/Pass kg/m³
11/1	1.0	5.0	130	4.5	37.0	
12/2	0.5	6.0	160	6.0	31.2	
12A/2	0.5	5.5	156	5.3		
13/1	1.0	5.0	124	5.0	37.2	33.9/32.9
14/2	1.0	2.5	171	5.7	38.4	33.6/35.2

*An average nominal density determined on several specimens tested for thermal performance.

Table 5. Initial thermal resistivity of 25-mm-thick specimens cut
from batches number 13 and 14 of the Base 88 foam system
(see: Bomberg and Kumaran, 1989).

Test Number	Location Density kg/m³	Number 1 r-Value m K/W	Location Density kg/m³	Number 2 r-Value m K/W
1	38.2	57.5	38.4	61.6
2		56.5	39.0	61.7
3	38.2	58.0	39.2	61.0
4	36.2	55.4	39.1	58.0
5	35.6	57.0	38.4	58.7
6	37.9	57.2	38.5	58.0
7	38.0	56.0	37.9	58.6
8	36.2	56.9	38.2	60.0
9		55.6	38.3	60.3
Average		56.7		59.8
Standard deviation		0.78 (1.4%)		1.49 (2.5%)

of SPF can be achieved in the field fabrication when the SPF contractor adheres to an appropriate QA program (see later text). In particular, use of two field measurements is recommended:

- time and maximum temperature rise (measured with a thermocouple placed on the first pass and sprayed over with a second layer of SPF or another gauge inserted into the SPF)
- core density (usually determined with a water displacement method)

Though high consistency and excellent long-term thermal performance of SPF can be achieved (Bomberg and Kumaran, 1989; Bomberg, 1993), this is not always the case in commercial practice. The quoted research showed that some HCFC blown SPF commercial products had thermal performance much worse than that of the traditional CFC-based sprayed foams.

Therefore, an industry-wide, quality assurance program must remain in the center of the SPF industry activities.

4. QUALITY ASSURANCE IS A KEY
TO THE SPF FIELD FABRICATION

The SPF industry [Spray Polyurethane Foam Division (SPFD) of the Society of Plastics Industries Inc. and Canadian Urethane Foam Contractors Association (CUFCA)] enforce industry-wide quality programs (see later text). This collection of articles will focus on relating these quality management programs to ensure long-term field performance of the SPF products and durability of building envelopes with SPF.

Long-term performance is seldom addressed in the traditional construction process. A traditional, "low bid" award system for selecting building contractors, favors the lowest initial cost. The fallacy of the low bid system is based on the assumption that since each bid must satisfy all technical requirements, long-term performance of all proposed systems will be warranted, at least above the "specified" minimum performance level. Such an assumption could only be justified if a specification was written in technical terms defining the actual field performance. Usually, however, this is not the case.

This research monograph introduces performance-based quality assurance (PBQA), a management system that stretches from a drum of chemicals through the specifier[1] to the contractor who fabricates the foam. Efficiency of every step in this quality management system will be judged by the result, i.e., the field performance of the construction component. This monograph expands the contractor training programs with methodology for predicting long-term performance of SPF developed during the CFC replacement program, operated in Canada during 1988–1994. Thus, the techniques used to predict durability and long-term field performance of the SPF systems have already been developed and verified, and the controls for the field fabrication of sprayed polyurethane foams are already in place.[2]

Alumbaugh (1993) and Bomberg (1993) underlined that the SPF contractors (field fabricators) must demand from each manufacturer (system house) technical data to ensure that the SPF system was adequately optimized with regard to its long-term thermal performance. This monograph assists SPF contractors by providing necessary test procedures for evaluation of the SPF systems and their testing during field fabrication. This monograph will also assist architects and designers by providing the design principles and construction details of walls and roofs with SPF. Designers must also pay attention to economic considerations associated with maintenance and repair of existing structures.

To highlight some issues, Figure 2 shows a general maintenance and repair model. A deterioration process reduces the performance level at the rate that is quantified by the damage function (Allen and Bomberg, 1997). The first case, A-scheme, shows that the structure may avoid major repairs if the preventive maintenance has been applied frequently enough. The second case, B-scheme, shows what happens when the structure is used with repairs of in-

---

[1] An architect or technical assistant who writes technical specifications for selected products.
[2] Experience with the company-wide QA model has been generated by Foam Enterprises Inc., the co-sponsor of this monograph.

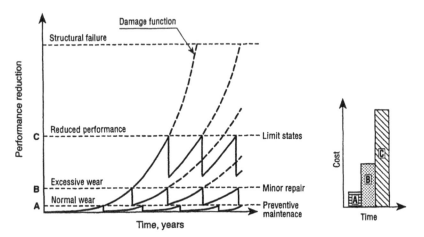

FIGURE 2. A model of maintenance and repair.

sufficient frequency. The total cost of repairs (a sum of all repairs and maintenance costs) is higher, yet the structure continues to perform well. Finally,
C-scheme shows a case of neglect, when repairs are made only at the start of
major failures. In this case, the repairs are made too late to restore the structure
to normal wear level. The structure will continue to function, however, at a
reduced level of performance.

Figure 2 showed that the pattern of maintenance and repairs affects the
long-term cost of the structure. One often advocates using a life-cycle-
cost (LCC) approach. The main drawback of the LCC approach is, however, its strong dependence on the estimated period of service life. In practice, when designing the structure with consideration to its long-term
performance, one selects, for the durability analysis, a reference period of a
typical return on investment, such as ten or fifteen years. This analysis
verifies if materials are durable for the required period of service under
specified climatic and service conditions. This verification can be performed with the limit states design method (see: Bomberg and Allen,
1996; Allen and Bomberg, 1997).

Thus, convincing architects and designers to perform a cost-benefit analysis, which includes the maintenance and repair considerations, is the next,
most important concern of the SPF industry. In doing so, the SPF industry
would harmonize product development with construction needs.

5. PERFORMANCE OBJECTIVES DEFINE THE SPF PRODUCTS

The performance-based quality assurance (PBQA) system proposed in this monograph, implies that one should first define performance objectives and then examine the best choice in the existing foam technology to satisfy these objectives.

Heat, air, and moisture aspects of building envelope performance are inseparable phenomena. Each phenomenon influences the others, but each is also influenced by the selection of materials within the building envelope. Often we simplify the design process by ascribing control of each phenomenon to a particular material. The thermal insulation is to control heat transfer and the air barrier is to control air leakage. Likewise, to eliminate ingress of moisture to materials, we use the rain screen and the vapor barrier.

However, each of these materials may perform different functions and influence several aspects of the overall performance. For instance, by controlling air leakage, the air barrier provides an effective moisture control. Similarly, by increasing temperature in the wall cavity in cold climates, an external insulating sheathing reduces the intensity of vapor condensation in the wall.

A holistic approach to building envelope design is not new, though a wide support for performance-based[3] codes is quite recent in North America. This approach brings, however, new demands for defining materials in the context of their contribution to the system performance.

With a view to their field performance, all sprayed, poured, and frothed polyurethane foams may be classified as follows:

1. Super high density (SHD-type)
2. High density (HD-type)
3. Medium density (MD-type)
4. Low density (LD-type)
5. Open-cell foam (OCF-type)
6. Bead-applied sealant foam (BSF-type)

The super high density (SHD-type) SPF is defined by the minimum compressive strength of 380 kPa (55 psi). In practice, it corresponds to type IV of the classification system used by the ASTM Specification C 1029 -90. The strength requirement will lead to a minimum foam core density[4] of 56 kg/m^3 (3.5 lb/ft^3).

The high density (HD-type) SPF is defined by the minimum compressive strength of 276 kPa (40 psi), leading to a minimum core density about 45 kg/m^3 (2.8 lb/ft^3). This type is identical to type III of the ASTM C 1029 -90 classification.

---

[3]Canadian Codes Center favors another term, namely "objective-based codes," which comes closer to describing "a set of code requirements which are based upon explicitly stated objectives."

[4]The limits postulated here are based on experience with HCFC-141b blown foams.

The SHD-type and HD-type foams are used primarily in roofing applications with a minimum 40 mm (1.5 inch) thickness. These spray polyurethane foams are strong, durable, and designed to satisfy requirements of roof traffic, forces of wind uplift, effects of rain, heat or extreme cold acting on the roof. The choice of one or the other type depends on the climate and service conditions, and on the intensity of traffic in particular.

The medium density (MD-type) and the low density (LD-type) foams are used in masonry or frame (wood or steel) walls. Though SPF in these applications is not directly exposed to the outdoor environment, it must withstand wind loads and structural movements. All loads, as well as environmental (hygric and thermal) stresses acting on these foams, are less severe than those acting on the roofing foams.

A minimum compressive strength of 173 kPa (25 psi) is required for the MD-type of SPF, resulting in the minimum core density of 37 kg/m^3 (2.3 lb/ft^3). The MD-type SPF corresponds to type II in the ASTM C 1029 -90 classification.

The LD-type is used in the building envelope only when structural movements and loads are negligible. A minimum compressive strength of 104 kPa (15 psi) is required for the LD-type of SPF. This compressive strength requirement results in a minimum density of 20 kg/m^3 (1.3 lb/ft^3) for predominantly closed-cell foams.

The closed-cell, LD-type SPF corresponds to type I in the ASTM C 1029-90 classification. The LD-type SPFs are designed to be more vapor-permeable (typically 20% to 30% more than MD-type), causing the hygric stresses acting on LD-type SPF to be much smaller than those acting on MD-type foams. However, in contrast to the MD-type foams, the LD-type foams do not qualify as a vapor retarder (see Chapter 2, Section 3 of this monograph).

As discussed later, the compressive strength is different in rise and normal directions. This difference is particularly large for polyurethane (either one-component or two-component) foams delivered in pressurized containers, where the compressive strength perpendicular to the foam rise may be as low as 50% of the compressive strength in rise direction. Therefore, for these LD-type foams, the minimum density would be 27 kg/m^3 (1.7 lb/ft^3).

The minimum recommended thickness for MD-type and LD-type SPF is 50 mm (2.0 inch) in masonry and 75 mm (3.0 inch) for the 89-mm-thick cavity of a frame wall (2 × 4 inch wood studs). This implies a 10–14 mm (nominal 1/2 inch) gap between the foam and inner gypsum layer (see Section 9.2 for further discussion).

Bead-applied sealant foam, called BSF is the last type of SPF. The name indicates that the foam is applied as a bead. It could be a bead of one-component or two-component foam, either chemically cured or moisture cured. There are many different products in the BSF-type category. Some of them (both two- and one-component foams with 90% of closed cell) have the compressive strength and density at the borderline of LD-type or higher. BSF-type foams have the thermal efficiency similar to that of expanded polystyrene. In effect, thermal resistance of 25 mm (1 ft) thick bead would range from 0.6 to 0.7 (m² K)/W or in IP units from 3.6 to 4.0 (ft² hr °F)/(Btu).

The properties of BSF-type foams are as follows:

- The density varies between 27 kg/m³ and 50 kg/m³ (1.7 lb/ft³–3.1 lb/ft³).
- The fraction of closed-cell ranges between 60% and 90 %.
- The compressive strength varies between 40 kPa and 80 kPa (6 and 12 psi). (Some commercial foam kits, however, contain foams with density as low as 16 kg/m³ (1 lb/ft³) and unspecified mechanical performance.)

BSF-type is usually applied when gaps or openings do not exceed 75 mm (3 in). Application of caulking compounds is not recommended when cracks are as small as 6 mm, then, BSF-type foam may be the best means of sealing the penetration. To dispense the foam, the can must be kept upside down, otherwise the propellant gas (typically nitrogen) will escape without delivering the foam.

Table 6 defines each SPF type by a requirement of the minimum compressive strength and summarizes the compressive strength/density relations for the six SPF types. The six SPF types listed in Table 6 are used in building envelopes as thermal insulation, roofing, air barriers, and penetration sealing. Table 7 illustrates other residential usage of SPF, primarily to control air leakage.

Existing on the market since the 1970's, one-component moisture-cured and two-component frothed-foams have found many applications including

Table 6. A minimum compressive strength and core density of each SPF type.

Types of SPF	Main Use	Property	SI Units	IP Units
SHD—super high density foam	Roofs	Compressive strength	380 kPa	55 psi
		Core density	56 kg/m³	3.5 lb/ft³
HD—high density foam	Roofs	Compressive strength	280 kPa	40 psi
		Core density	45 kg/m³	2.8 lb/ft³
MD—medium density foam	Walls	Compressive strength	170 kPa	25 psi
		Core density	37 kg/m³	2.3 lb/ft³
LD—low density machine applied pressurized container	Walls	Compressive strength	100 kPa	15 psi
		Core density	20 kg/m³	1.3 lb/ft³
		Core density	27 kg/m³	1.7 lb/ft³
BSF—bead applied foam sealant	Air sealing	Compressive strength	35 kPa	5 psi
		Min. density	16 kg/m³	1.0 lb/ft³
OCF—open cell foam	Air sealing	Compressive strength	N/A	N/A
		Min. density	8 kg/m³	0.5 lb/ft³

*Table 7. Residential usage of SPF for air leakage control
(Braun, Hansen and Woods, 1995).*

	One-Component (BSF)	Two-Component (LD- or OCF-types)
Basement and crawl spaces		
Sillplate	√	√
Headers	√	√
Pipe penetrations–air conditioning, water, gas, oil	√	
Duct penetrations and shafts	√	√
Wire penetration—TV, antenna	√	
Conduit penetrations—electrical panel	√	
Hose bib	√	
Dryer vent	√	
Plumbing stacks and shafts	√	√
Windows and surrounds	√	
Floor/wall junction	√	
Wall cracks	√	
Doors—cold room and exterior	√	
Sump pump holes	√	
Behind bath tubs		√
Crawl space insulation and seal		√
Living Space		
Baseboards—interior and exterior walls	√	
Windows and trim	√	
Electrical receptacle and switches—interior and exterior walls	√	
Plumbing penetrations and hatches	√	
Wire penetrations	√	
Heating vent perimeters	√	
Cold air return ducting		√
Door and framing—exterior, patio and pocket	√	
Interior door trim—cupboards, etc.	√	
Exhaust fan perimeters	√	
Baseboard heaters	√	
Range hoods	√	
Attic		
Attic hatch	√	
Ducting	√	√
Plumbing stacks	√	√
Wire penetrations	√	
Headers	√	√
Recessed ceilings		√
Light fixtures		√
Knee walls		√
Pipe penetrations	√	
Pot lights—boxed with drywall	√	
Party walls	√	√

sealing around doors and windows, joining insulating panels, and draft-proofing around cables and pipes. Although these foams alone do not have better characteristics than other materials used for filling the holes, the continuous and well-bonded seal around the penetration improves the acoustical performance of the construction element.

Table 6 introduced six SPF types and Table 7 showed more than forty places where either one- or two-component foams could be used for sealing cracks and openings. It is evident that to select an appropriate SPF product, one must know more about factors affecting field performance of SPF.

CHAPTER 2

Factors Affecting Performance
of the Foam

THIS CHAPTER EXAMINES several aspects of SPF performance such as thermal resistance, transport and accumulation of moisture, integrity, and dimensional stability of the foam under conditions of varying temperature and humidity.

1. THE AGING PROCESS (THERMAL DRIFT)

The aging process can be explained using the distributed parameters continuum (DIPAC) model developed and verified experimentally by the National Research Council of Canada (Bomberg and Kumaran, 1995). The DIPAC model illustrates the relative significance of different aging mechanisms. Figure 3 shows thermal resistivity (inverse of thermal conductivity coefficient) versus aging time.

Four curves are shown in Figure 3. Curve 1 shows the aging of a 25 mm- (1 inch) thick SPF specimen, fully encapsulated on all sides. The encapsulation prevents entry of air into the foam, but has no effect on the redistribution of the BA within the encapsulated foam. Part of the BA enters and saturates the polymer matrix, reducing the concentration of the BA in cell gas. This, however, does not change the thermal performance of the fully encapsulated specimen. The thermal conductivity of the cell gas does not depend on the pressure of the gas [as long as the gas pressure does not fall below 0.01 atmosphere (Tsederberg, 1965)]. The thermal conductivity (k-factor) does not change as long as air has not entered the cells of the foam, despite the change in pressure caused by the cell gas redistribution. For example, foams with impermeable sheet metal facings demonstrate high thermal performance for extended periods (Baumann, 1982). This is true even if the thermal efficiency of the BA is low, e.g., carbon dioxide.

Curve 2 shown in Figure 3 relates to the hypothetical aging of the same specimen when only air is allowed to enter the foam. In this computer simulation, the BA redistribution is eliminated by using zero values for the effective diffusion and solubility coefficients of the BA; the effective diffusion coefficients for oxygen and nitrogen are taken from Bomberg and Kumaran (1989).

17

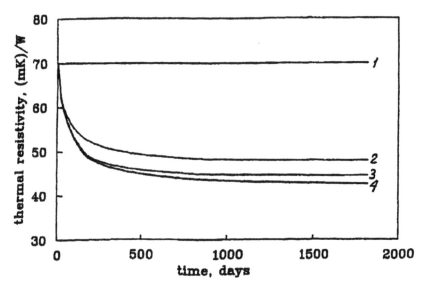

FIGURE 3. Hypothetical curves of thermal resistivity versus aging time for a 25 mm-thick, typical sprayed polyurethane foam (SPF) specimen.

Curve 3 in Figure 3 shows the increase in the aging process when the effects of BA solubility in the polymeric matrix are included in the calculation. Finally, Curve 4 adds one more mechanism of the aging process, namely that of the BA outward diffusion. The difference between Curve 3 and Curve 4 is small. This shows that the thermal drift is primarily caused by the ingress of air into the foam and the solubility of the BA. The BA diffusion affects the aging process much less than expected.

Twenty years ago, thermal drift was associated only with the outward diffusion of the BA. The research of Brandreth and Ingersoll (1980, 1981) and Ostrogorsky and Glicksman (1986) highlighted the importance of air ingress into the foam. On the other hand, CFC solubility was examined by Steinle (1971). He observed that during the first 6 to 12 month period as much as 48% to 58% of the total CFC-11 content was dissolved in the cell walls of the polyurethane foam. The significance of this information was then neither understood nor appreciated. The consensus in the industry was that the BA solubility could be neglected and the total difference between the initial and the long-term thermal performance could be ascribed to the diffusion processes.

The understanding of aging was improved during the research on CFC replacement, mainly because to establish a basis for evaluating foams with alternative BAs, a better insight into the aging phenomena was necessary. A joint project of the Canadian Plastics Industry Association and the National Research Council of Canada (CPIA/NRCC) was initiated. A generic, CFC-11 blown SPF was developed as a benchmark material. The generic SPF was called Base 88. This technology was subsequently carried through the BA replacement program into SPF products manufactured and fabricated with partly halogenated hydrochlorofluorocarbons (HCFCs). Table 8 lists the composition of this SPF system.

Using 30 parts per weight of BA (BA) in the polyol and assuming a 3 to 4% BA loss during foam spraying (discussed below), the initial BA content is about 11% of the foam weight. During spraying, an exothermic reaction increases the foam temperature to a maximum of 130° to 160°C. BA gas pressure, which is equal to 1 to 1.05 atmospheres at the peak temperature, is then reduced to about 0.75 atmosphere when the foam cools to the ambient temperature.

Subsequently, part of the BA enters into and saturates the polymer matrix, reducing the BA pressure in the cell gas. A nondestructive technique was developed (Kumaran et al., 1989) to determine the partial pressure of the BA inside the cells of the foam (Ascough et al., 1991). The measurements were performed on specimens encapsulated with impermeable coating to "freeze" the aging process. These measurements indicated that the CFC-11 pressure in Base 88 foam decreased to about 0.55 atmosphere within a few weeks and to 0.43 to 0.44 atmosphere after seven to nine months. Because these measurements were made on fully encapsulated specimens, the reduction of BA pressure cannot be ascribed to the outward diffusion, but implies a significant solubility of the BA in the polymer matrix. Table 9 summarizes these findings (Bomberg et al., 1991), indicating that the solubility of both HCFCs are larger than that of CFC.

Other ten- to twenty-year old polyurethane samples, retrieved from field exposures and/or NRCC laboratory storage, corroborate results shown in

Table 8. The composition of the generic spray polyurethane, Base 88.

Component	Percent by Weight
Aromatic amino-polyol 1	17.0
Aromatic amino-polyol 2	17.0
Polyester polyol	10.0
Triethanolamine	13.0
Glycerin	5.5
Flame retardant	5.0
Silicone surfactant	1.0
Lead catalyst	0.3
Amine catalyst	1.0
CFC-11	30.0

Table 9. BA pressure after eight to ten months of aging measured on encapsulated SPF specimens (Bomberg et al., 1991).

Encapsulated after Day	Pressure of the BA in Atmospheres		
	CFC-11	HCFC-123	HCFC-141b
10	0.43	0.35	0.33
16	0.44	0.35	0.33
21	0.43	0.35	0.33
26	0.44	—	0.35

Table 9. Some rigid polyurethane and SPF products show a high retention of CFC-11 in the cell gas even after many years. For example, SPF sample coded 293-173 (see Bomberg and Brandreth, 1990) showed 70% of the CFC-11 in the cell gas after 11 years of laboratory storage. Other foams show much worse retention of CFC in the cell gas. The SPF sample coded 396-89 showed that the reduction to 54% of the CFC-11 in the cell gas took place within a few weeks of manufacture.

A significant variation in BA solubility has also been observed in the generic foam, Base 88. Samples taken from six different batches, from a few weeks to a few months old, were tested in the same manner as discussed by Bomberg (1980). The total CFC content ranged from 10% to 12.2% of the foam weight (Bomberg and Kumaran, 1989). However, the distribution of the BA between the cell gas and the polymer showed large variations. Two batches of Base 88, batch 13 (fabricated at location 1 using isocyanate 1) and batch 14 (fabricated at location 2 using isocyanate 2), were compared. The CFC-11 solubility was very high (60%) for batch 13 and very low (22%) for batch 14. There was also a 5.6% difference in the mean values of the initial thermal resistance, and batch 14 displayed better thermal performance than batch 13 over the whole aging period (Bomberg and Kumaran, 1989).

The significance of solubility was also noted by Booth and Lee (1986) and Wiedermann et al. (1987). Gaarenstroom et al. (1989) showed that the actual emission of CFC-11 from the foams was much smaller than previously assumed. After twenty-one years of aging, the total amounts of CFC-11 in the foam core in the outer layers of the foam were 87% and 86% respectively. When correcting for loss that occurs during production, one may conclude that very little of the BA actually escapes from the sample. The reduction of thermal efficiency, previously ascribed to the loss of BA, can now be explained in a more plausible way.

The aging process is often presented in a graphic form to reduce the scatter in the measured values and to extract the long-term trend. Since aging involves diffusion processes, the thermal resistivity (the inverse of the k-factor) may be related to the aging time using either the square root of time or the logarithm of time. Each of these two graphical forms distinguishes between two stages of the aging process:

- the primary aging stage (thermal drift stage)
- the secondary aging stage (the stabilized k-factor stage or plateau stage)

The first stage shows a rapid reduction in the thermal resistivity of the foam, related to changes in the cell gas composition, that are caused by carbon dioxide leaving the cells and air components entering the cells. The second stage is characterized by slow changes in the cell gas composition because the air ingress is completed and the outward diffusion of the BA changes the cell gas composition at a much slower rate.

To permit normalization of the measured properties, Figure 4 uses a logarithmic time-scale. While the thermal resistivity depends on both time-dependent and time-independent factors, the normalized thermal resistivity

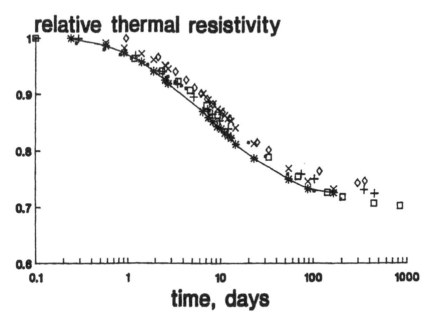

FIGURE 4. Normalized thermal resistivity (the actual thermal resistivity in relation to the initial value) against the logarithm of time for six Base 88 batches produced at two locations with different isocyanate products. Reprinted with permission of NRCC.

(thermal resistivity divided by its initial value) is only affected by time-dependent factors. In effect, the normalized aging curve, such as that shown in Figure 4, becomes an intrinsic material characteristic.

Figure 4 shows the lowest and highest aging curves for thin layers cut from six batches of Base 88. Half of the batches were fabricated at location 1 using isocyanate 1, the other half were fabricated at location 2 using isocyanate 2 (Bomberg and Kumaran, 1989).

By virtue of the relative character of the aging curves (Figure 4), the effect of time on the thermal resistance is related only to the time-dependent processes of diffusion and solubility. Figure 4 indicates a high degree of consistency of the solubility and diffusion characteristics of the six batches examined. While the differences in the composition of the cell gas affect thermal resistance of separate batches, the identical polymeric composition of all the foams resulted in a similar gas diffusion coefficient. Indeed, by using the relative thermal resistivity concept, the same diffusion coefficient was also observed in two apparently different types of polyurethane and one product manufactured in two different plants (Edgecombe, 1989). Thus, under specific conditions one may use the normalized aging curve for calculation of the effective diffusion coefficient. These conditions include only one gas being diffused through the foam only and that, during the test, the change in the amount of dissolved BA is insignificant. Such a method has been proposed by Booth and Holstein (1993).

Improved understanding of the thermal drift phenomena (Booth, 1991; Glicksman et al., 1991) provided new opportunities. CPIA/NRCC research developed the HCFC-blown SPF with a long-term thermal performance equal to CFC-11-blown foam (Kumaran and Bomberg, 1989, 1990).

2. THE DESIGN THERMAL RESISTANCE

Traditionally, in North America, thermal resistance of the product is determined on a dry piece of material with prescribed dimensions, e.g., 25 mm- (1 inch) thick, at a standard temperature of 24°C (75°F). However, thermal properties determined in such an arbitrary manner seldom describe material performance under field conditions.

A number of environmental effects will alter field performance of the thermal insulation. Thermal performance of low-density, glass fiber insulating products is often affected by air flows. Cellulose materials are often affected by moisture absorbed under service conditions. Gas-filled cellular plastics are subject to aging and their performance varies with time. Field performance of thermal insulation may, therefore, be quite different from that measured under standard laboratory conditions.

As a result, many European countries use concepts of declared value and design value (ISO, 1991). The declared value is the expected value of the thermal characteristic of a building material or product assessed through data measured at a reference temperature and thickness and stated with a given confidence level. The design value is the value of the thermal characteristic of a building material or product in a condition that can be considered typical in buildings. The design value may depend on climate and use conditions.

An ISO draft, "Methods for determining declared and design thermal values" discusses the procedures to derive the declared and design values, and introduces effects of temperature, moisture, age of the material, thickness, and other factors. Some details in this ISO draft may still be the subject of discussion, but the use of design thermal properties helps users in selecting materials with better understanding of their field performance. Conversely, the current North American practice is based on laboratory data obtained on dry and fresh specimens. The relationship between this data and the true field performance of the material is often vague and is mainly established through the individual experience or tradition.

Thermal resistivity of the material is also used for estimating the thickness of insulation to achieve the thermal resistance needed for the prescribed heating or cooling loads. Because the thermal resistance of gas-filled foam varies with time, the heat loss (or gain) calculations should be based on thermal properties averaged over the service life. In practical terms, a reference point in the aging period should be selected which yields a thermal resistance equal to the time-averaged value.

Di Lenardo and Bomberg (1995) performed an economic/technical analysis to establish the error introduced by an "assumed" reference aging period. This analysis involved five different, cellular plastic insulating products, three Canadian climates, and two heating fuels. A number of variables were identical to those used for development of the 1995 Canadian Energy Code. These included the cost of construction of the wall assemblies, and the cost of insulation alternatives as well as economic assumptions such as future interest rates, inflation rate, and fuel escalation rate.

This analysis indicated that the same reference time may be used for a wide range of products that have thermal performance varying during the service time and design (service) life between fifteen and twenty-five years. The reference time could be as short as two years, or as long as five years, and no more than 2% of difference would be created for thermal insulating products with thickness between 20 mm and 90 mm. Note that the reference time of five years was previously postulated by a group of experts in a position paper (Kabayama, 1987); and the reference period of two years was postulated by California state ruling for evaluation gas-filled thermal insulations.

The research performed by Di Lenardo and Bomberg (1995) highlighted that using long-term thermal resistance (LTTR) instead of the initial thermal resistance is of paramount significance. This research also pointed out that whether the reference period is two or five years has a much smaller effect than often thought. Since the practical investment considerations relate to return on investment in fourteen to eighteen year period (no less than ten), either 10- or 15-year aging period were discussed. The 10-year period was recently discussed by the Polyisocyanurate Manufacturers Association (PIMA) and the 15-year aging period was agreed by the Canadian Plastics Industry Association (CPIA). The corresponding reference periods[5] are 3.5 or 5 years respectively. Either period is sufficient for evaluation of the long-term performance of cellular plastics, though the CPIA recommendation is closer to those of ISO, CEN and the general consensus on LTTR.

3. TRANSPORT AND ACCUMULATION OF MOISTURE IN THE FOAM

Water vapor transport through rigid polyurethane and polyisocyanurate foams was investigated by Schwartz et al. (1989) under isothermal conditions as well as in the presence of thermal gradients.

3.1 Moisture Transmission through the Foam Core Material

One test method, the dry cup procedure of ASTM E 96, is a standard used for testing water vapor performance through many construction materials and membranes. A test specimen is placed between two compartments, one with controlled temperature and humidity and another where a desiccant maintains humidity at near zero conditions. As the same temperature is applied to both sides of the specimen, this is an isothermal test.

Another isothermal test method used in this research was a modified cup method, which does not require control of the ambient humidity. Furthermore, the modified cup method allows the determination of changes in moisture content of the specimen. Using this method, it was conclusively shown that no moisture accumulates in either polyurethane or polyisocyanurate specimens during isothermal water vapor transport. This was not the case during the non-isothermal WVT testing.

The non-isothermal setup of Schwartz et al. (1989) used a Heat Flow Meter apparatus which maintained a temperature of 50°C on the hot plate and 5°C on the cold plate. A water container, made of Plexiglas, was placed on the hot plate. On the cold side of the specimen, there was either a container with a desiccant or

---

[5]A reference period is the time when thermal resistance of the foam becomes equal to an average for the considered period. The reference period is important because it allows comparing the measured, actual R-values with those predicted. The relation between the long-term aging period and the reference period is approximately 1 to 3.

a vapor barrier. The first set up allowed a flow of moisture through the foam; the second prevented such a transport. The specimens were periodically removed from the assembly in order to measure their moisture gains (by weight) and to determine the distribution of moisture in the specimen (with a gamma-ray spectrometer described by Kumaran and Bomberg, 1985).

Table 10 lists the water vapor transmission measured on polyurethane (SPF) and polyisocyanurate (PIR) specimens. Standard error, estimated from the least square analysis, is shown in the parentheses. It was observed that the magnitude of the water vapor permeance increases with temperature (above 20°C). It was also observed that the moisture accumulation inside the foam did not change the rate of vapor transport through the SPF and PIR foam. While there was variation between results of different measurements, there appears to be no systematic difference between the results obtained from isothermal test methods (dry cup and modified cup methods) and those obtained from the thermal gradient method.

While the rate of water vapor transmission appears similar with and without the presence of a thermal gradient, this is not the case with moisture absorption. Contrary to the isothermal process, the moisture transport in the presence of a thermal gradient acting in the same direction as the vapor pressure gradient results in the accumulation of large quantities of water inside the foam. This phenomenon can only be partly attributed to the temperature dependence of the permeance (see Figure 5).

Figure 6 shows moisture gains by the SPF and PIR specimens exposed to the desiccant on the cold side. Figure 7 shows moisture gains when the specimens were covered by impermeable films on the cold side. In both cases, the moisture gain stabilizes after some time. However, the period and the degree of moisture accumulation are affected by the conditions on the cold side. PIR specimens shown in Figure 6 show smaller gains than SPF specimens. Their moisture gain stabilizes after approximately forty days in the experiment with desiccant, and seventy after days when the cold side is covered with an imper-

Table 10. The water vapor permeance, $\mu \times 10^7$, in g/(m²·s·Pa), for SPF and polyisocyanurate (PIR) foams. From Schwartz et al. (1989).

Temperature °C	Test Method	SPF, $\mu \times 10^7$ g/(m²·s·Pa)	PIR, $\mu \times 10^7$ g/(m²·s·Pa)
10.0	Modified cup	1.14 (2.2)	1.02 (1.1)
21.5	Dry cup	1.12 (0.7)	1.03 (2.3)
21.5	Modified cup	1.16 (1.2)	1.01 (1.0)
Approx. 27.5	Thermal gradient	1.80 (2.1)	1.31 (3.3)
32.0	Modified cup	2.09 (0.1)	1.62 (5.3)
51.0	Modified cup	—	2.66 (5.0)

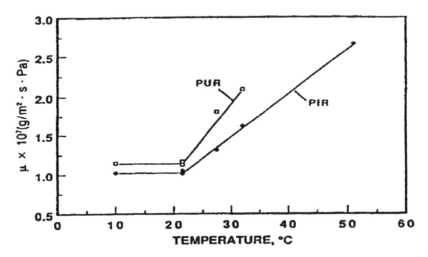

FIGURE 5. Temperature dependence of water vapor permeance, $\mu \times 10^7$, in g/(m²·s·Pa), for SPF and PIR foams with density about 23 kg/m³. Reprinted with permission of NRCC.

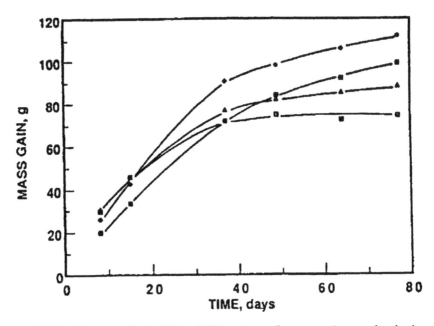

FIGURE 6. Mass gained by the SPF and PIR specimens (lower curves) exposed to the desiccant on the cold side. Reprinted with permission of NRCC.

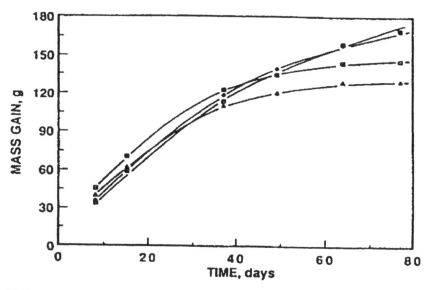

FIGURE 7. Mass gained by the SPF and PIR specimens (lower curves) covered by an impermeable film on the cold side. Reprinted with permission of NRCC.

meable film. SPF specimens appear to exhibit similar behavior, but it takes a slightly longer period to reach their stabilized moisture content, and their moisture gains are somewhat higher.

Figure 8(a) shows the distribution of moisture in the SPF specimens at two stages of the experiment performed with the desiccant on the cold side. Figure 9(a) shows the moisture distributions obtained at the same time on SPF specimens when the cold side was covered with an impermeable film. A periodic gamma-ray scanning throughout foam specimens exposed to a thermal gradient showed that significant quantities of moisture were deposited in the middle of the material. Moisture build up in the specimen continued until the equilibrium was reached between the vapor flowing in and out of the specimen. Figure 9 shows a gradient of moisture content at the cold side, although this gradient is much smaller than one shown in Figure 8. There is also a difference in the slope of the ascending part of the moisture distributions shown in Figures 8 and 9.

The information on moisture accumulation in SPF presented in Figures 5 through 9 can be generalized as follows. If the vapor transport occurs under isothermal conditions, there is no significant accumulation of moisture. If the process occurs in the presence of a thermal gradient, moisture will be accumulated in the middle of the foam. One may also observe that the accumulation of moisture when the cold surface was covered with an impermeable film was much higher than when the surface was exposed to air. Yet, accumulation of moisture is significant in both cases.

Figures 6 through 9 show that accumulation of moisture under conditions of thermal gradient was significant. This contradicts a widespread misconception about moisture performance of closed cell cellular plastics. It is often assumed that if a cellular plastic (foam) does not absorb moisture when immersed in water, then it also will not absorb moisture under typical service conditions. This assumption is evidently false. The water ingress is stopped because 94% to 98% of the void space is enclosed by continuous cell membranes. Water vapor, however, as shown in Figures 6 through 9 can easily migrate through these membranes and condense inside the foam cells.

Different foams may accumulate moisture to varying degree, depending on the interaction of thermal and humidity gradients. Figure 10 compares mois-

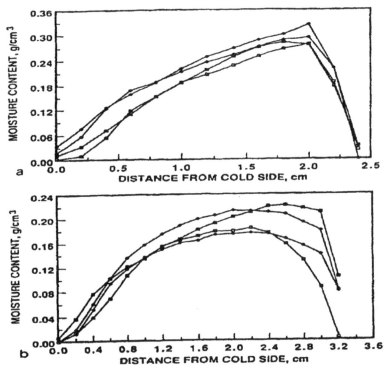

FIGURE 8. Moisture distribution in the SPF and PIR [Figure 8(b), lower set of curves] specimens exposed to the desiccant on the cold side after thirty-seven and sixty-four days of exposure. Reprinted with permission of NRCC.

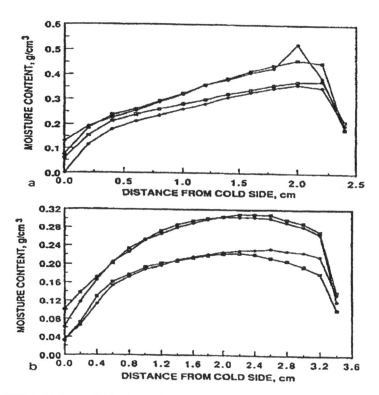

FIGURE 9. Moisture distribution in the SPF specimens and PIR [Figure 9(b), lower set of curves] covered by an impermeable film on cold side after 37 and 64 days of exposure. Reprinted with permission of NRCC.

ture absorption of two types of expanded polystyrene and polyurethane. Low-density expanded polystyrene (15 kg/m³) is denoted as material 1, medium density polystyrene (21 kg/m³) as material 2, and polyurethane with density of 40 kg/m³ as material 3. To determine the amount of moisture accumulation in the foam, these three materials were tested under three different environments (Bomberg, 1983, Forgues, 1983). These test methods, however, measure the apparent moisture content, which includes the moisture attached to the material surface. To derive the true measure of moisture penetration into the foam, a correction for surface water content is required. Therefore, Figure 10 shows only the qualitative relationships obtained in these laboratory tests.

To understand those differences, one must realize that water contained within the cells of the foam always evaporates and diffuses toward the lower vapor pressure associated with the lower temperature. There, water vapor may or may not condense. If the vapor concentration exceeds the saturation concentration at the given pressure, the vapor will condense again. This reduces the vapor concentration in the cell gas and further enhances the diffusion process towards the cold side of the material. This phenomenon is called a thermal moisture drive (TMD).

The ranking order for moisture absorption obtained in tests with different boundary conditions varies. In the water immersion test (isothermal conditions), material 1 showed the highest apparent moisture content, material 2 was second, and material 3 was third. Freeze-thaw conditions produced different results. All materials showed higher moisture absorption, but material 2 showed the highest apparent moisture content, followed by material 1, and material 3. The ranking order was further altered when specimens were exposed to a constant temperature difference.

The TMD affects polyurethane foams much more than polystyrene foams. The amount of moisture absorbed by the polyurethane foam greatly exceeds that absorbed by EPS types 1 and 2, and the ranking order under a constant temperature gradient is 3, 2, 1.

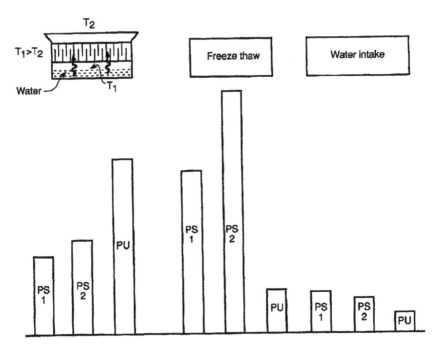

FIGURE 10. Moisture absorption of three types of cellular plastics under different thermal environments. Test 1 represents a constant gradient of temperature; test 2 = freezing and thawing cycles; test 3 = isothermal water immersion. Reprinted with permission of NRCC.

The example shown in Figure 10 illustrates the significance of different environmental conditions. The same SPF that absorbed insignificant amounts of moisture when immersed in water and collected a small amount of moisture when exposed to freeze-thaw conditions, absorbed large quantity of moisture under the effect of a constant thermal gradient.

In effect, when service conditions involve one-directional heat flow through most of the service time (e.g., cold storage), the use of SPF requires special design considerations.

3.2 Effect of Skins on Moisture Transmission through the Foam

The surface layer of the same SPF product exhibits different properties when sprayed on different substrates. Two substrates were examined, a smooth, non-capillary material (polyethylene) and a porous, hygroscopic material (plywood). While both foam layers exhibit the same character of aging when thermal performance is normalized, modifications in cell morphology caused by the effect of substrate were shown to affect thermal resistance of the foam (see Chapter 3, Section 1.4). Would these changes also affect the moisture performance of the foam?

To examine this issue, testing[6] was performed at NRCC with one SPF product applied to different substrates: an external-grade gypsum sheet and a concrete block wall. These tests included first measuring: WVT on the whole system (a composite of the foam and the substrate) and subsequently, measuring WVT on each of its components separately. Each test involved three specimens tested in accordance with ASTM test method E96. Table 11 shows the obtained results.

Table 11 highlights differences between the WVT of the whole system and the sum of its components. The resistance to vapor flow of the composite product (SPF and the exterior gypsum board) is slightly higher than the sum of values obtained on each of the system components tested individually. This difference indicates that the interface provided an additional resistance to vapor transmission. The resistance added by the interface of SPF and gypsum is not large—it is about 5×10^{-9} (Pa·s·m²)/kg, which corresponds to a material layer with a water vapor permeance of 3.6 perms.

When SPF was applied to the surface of the concrete block, the composite foam/block has a much higher resistance toward water vapor transmission than the sum of the values obtained on each of the components tested individually. The resistance added by the interface of the SPF layer and the masonry substrate is quite significant, and corresponds to a material layer with a water vapor resistance of approximately 16.8×10^{-9} (Pa·s·m²)/kg (permeance about 1 perm).

In effect, even a thin layer of the SPF applied to masonry block wall will perform the function of vapor retarder as it is defined by the majority of building codes in

---

[6]The sponsor of these tests was Demilec Inc., Boisbriand, Quebec.

Table 11. The average WVT measured with ASTM test method E96 on components and systems involving SPF. (Details-types, 1997).

Component or System	Co	Thickness (in)	WV Resistance (Pa·s·m²)/µg	Permeance ng/(Pa·s·m²)	Permeance Perm
External gypsum	A	12.6 (1/2)	0.56	1787	31.3
MD-type SPF	B	27.7 (1.0)	9.2	109	1.91
SPF on ext. gypsum	A+	39.9 (1.5)	14.7	68	1.19
Difference			4.9	204	3.6
Concrete block web	C	21.8 (0.8)	3.79	271	4.8
MD-type SPF	D	26.5 (1.0)	6.99	143	2.5
SPF on concr. block	C+	47.9 (1.8)	27.47	36.4	0.64
Difference			16.8	59.5	1.04

North America. Table 12 shows approximate results for water vapor permeance that can be used for design purposes for 50 mm (2 in) or 75 mm (3 in) thick layers of MD-type SPF. They are based on "skin effects" as shown in Table 11 and WVT characteristics of type II foam as listed in the ASTM C 1029 standard.

4. COMPARATIVE TESTING OF SPF DIMENSIONAL STABILITY

A test that examines dimensional changes of a foam placed in a hot and humid environment may be used for screening polymers and may assist in the design of stable foam systems.

4.1 Proposed Clarifications to ASTM D 2126 Standard Test

A laboratory (benchmark) test, though it may not represent performance under field conditions, may still provide valuable information for improving

Table 12. The average design values of water vapor permeance for SPF systems.

System	Total Thickness mm (in)	WV Resistance (Pa·s·m²)/µg	Permeance ng/(Pa·s·m²)	Permeance Perm
50 mm (2 in) SPF sprayed on external gypsum	63 (2 1/2)	19	52	0.91
75 mm (3 in) SPF sprayed on external gypsum	89 (3 1/2)	25.7	39	0.68
50 mm (2 in) SPF sprayed on concrete blocks	74 (2 7/8)	33.9	29.5	0.52
75 mm (3 in) SPF sprayed on concrete blocks	98 (3 7/8)	40.6	24.6	0.43

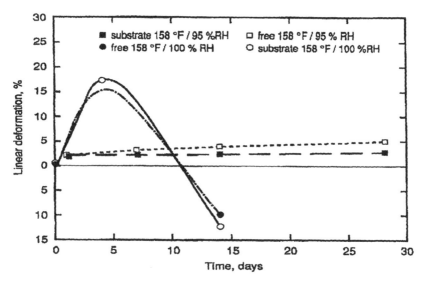

FIGURE 11. Comparative study of dimensional stability of the SPF tested in the form of free slabs and bonded to the waferboard substrate in two environments. The temperature was 70°C (158°F), but relative humidity was either below 95% or near 100% RH, respectively. Reprinted with permission of NRCC.

the design of a product. Such a test, however, must yield a high degree of precision and a unique interpretation. This does not appear to be the case with the current ASTM standard D 2126. This test method needs clarifications, as discussed below, in four areas.

The first concern is the 14% volume expansion limit allowed by this standard, which may mean many different things. This might be a 2.4% linear expansion in all directions or a 10% expansion in a single direction, 5% in another and some shrinkage in thickness.[7]

The second concern is the lack of clarity, this test does not explain whether vapor condensation (conditions near 100% RH) is to be attained or avoided. This means a poor reproducibility of the test. Figure 11 shows results of a comparative study of dimensional stability performed by four different laboratories. Relative humidity varied between 92% and 98% RH in three laboratories. A special device used by the fourth laboratory kept the relative humidity in the narrow range of 98% to 100% RH, i.e., surface condensation conditions (Report, 1987). All specimens were made during one fabrication process and applied to a waferboard substrate. Two patterns were used—half of the specimens were cut as free slabs and the other half were left attached to a

---

[7]We propose that, in addition to the volume change, the average thickness and the average linear change in two other dimensions (approximate as they may be) be reported.

waferboard substrate. All specimens had a SPF thickness of 25 mm (1 inch) and dimensions of 300 × 300 mm (12 × 12 inch).

As expected, there was a slight difference between dimensional changes measured on free foam-slabs and the specimens attached to the substrate. Nevertheless, the pattern of dimensional changes measured on free or attached specimens was similar, indicating that free slabs may be used for comparative testing. A dramatic change in dimensional stability occurred, however, when the foam was subjected to moisture condensation, as was the case with one laboratory. The foam showed rapid and strong expansion over four days of exposure, reaching 13% to 17% of the linear change followed by a significant shrinkage over the next 10 days, reaching minus 9% to minus 12%. When tested after fourteen days, these specimens showed such a low value of thermal resistivity that one could infer that the rupture of many cells caused a rapid ingress of air and loss of the BA.

SPF exposed to high humidity without the condensation of moisture showed a different behavior. These SPF specimens showed a slow and uninterrupted expansion throughout the testing period resulting in a 3% to 5% change over a four week period. The issue of moisture sensitivity was well recognized in the early stages of spray foam development (Hilado, 1967; Levy, 1966), but appears to be disregarded today.

To provide information suitable for the design of SPF products, ASTM D 2126 test method should require relative humidity in excess of 98% (i.e., surface condensation).

The third concern is that the correlation between results obtained from this laboratory test and field performance of the foams was lost when CFCs were replaced with HCFCs. Since the boiling pressure and solubility characteristics of the BA affect the development of cell-gas pressures, the introduction of HCFC into the foams changed the relation between exposure time and measured deformations. The traditional period of testing became insufficient to yield the most critical information and modifications were needed.

The fourth concern is more fundamental in nature. How can a test employing isothermal conditions be used to differentiate between poor and good SPF field performance, when the presence or absence of thermal gradient is known to change the character of dimensional changes? For instance, a significant difference between dimensional stability of SPF exposed to near condensation conditions ($\Phi > 98\%$ RH) was observed when testing was performed with and without thermal gradient. Significant specimen warping was observed when WVT was tested with the isothermal, extended cup method (relative humidity near 100% RH on one side and 0% RH on the other side). There was, however, no specimen warping when the thermal gradient was applied. The same moisture conditions were used in the non-isothermal test, namely high relative humidity (Φ > 99% RH) with temperatures of 35°C and 13°C on the side with desiccant (Φ near 0% RH).

The following section provides further information on the dimensional stability of foams exposed to cycling environmental conditions.

4.2 A Response to the Environmental Cycling of Foams

In the absence of any generally accepted procedure on accelerated weathering of thermal insulation, a standard procedure for testing the durability of double-sealed glazing units was used for exposure of different foam products (Schwatz, 1991). Samples from five extruded polystyrene (XPS), five SPF, and two phenolic (PhR) foams were used in a pilot project comparing dimensional stability of foams under environmental cycling.

Three, 300 mm-square specimens from each material (25 to 37 mm thick) were placed as dividers between two environments. One side of the specimen was exposed to room conditions; the other side was exposed to one-sided, 4 hours weather cycles. During the weather cycling, the chamber temperature varied between −29°C and 50°C, and during the cooling portion of the cycle, when the temperature reached 25°C, the specimen received a water spray lasting for five minutes. The exposure lasted for 28 days (i.e., 168 cycles).

Substantial moisture absorption of phenolic foams may be noted (Table 13). SPF showed smaller increase in weight, while XPS appeared to be almost unaffected by the water sprayed under freeze-thaw conditions.

Standard tests for dimensional stability were also performed in a manner

Table 13. Changes (percent) introduced by 168 freeze-thaw cycles.

Sample	Length	Width	Thickness	Weight
XPS 2	−0.20	−0.40	−0.87	−1.17
XPS 3	−0.12	−0.24	−0.49	−0.43
XPS 4	−0.06	−0.32	0.03	−0.02
XPS 5	1.30	−0.27	0.50	0.85
XPS 6	−0.11	0.24	−0.13	−0.16
PhR 1	0.69	0.33	—*	110.7
PhR 1	0.05	1.07	—*	383.3
PhR 2	0.02	0.68	—*	121.7
PhR 2	−0.23	−0.52	—*	104.3
SPF 1	0.12	0.46	4.38	1.17
SPF 2	0.27	0.39	4.33	3.16
SPF 3	−0.04	0.80	4.26	2.33
SPF 4	0.06	0.66	5.27	4.04
SPF 5	0.63	0.45	6.41	5.21

*Substantial warping was observed.

consistent with the Canadian material specifications: exposure at 80°C and at −29°C for PhR; exposure at 70°C for XPS; and exposure at 100°C for SPF. This study was to determine whether the material response to environmental cycling was similar or different from that obtained during the dimensional stability tests requested by material specifications.

The dimensional stability of XPS was nearly the same in both tests. The phenolic foam tested at an elevated temperature showed linear shrinkage of 1% to 3% in all dimensions as opposed to the swelling and warping during the freeze-thaw cycling (Table 13). The latter changes are likely caused by moisture in the material.

The SPF products showed small changes when tested at −29°C, and significant changes at elevated temperatures (100°C for 28 days). All specimens exposed to the elevated temperature test shrank 0.1 to 2.0% in thickness and expanded between 4 and 10% in the other two dimensions. Some specimens also showed significant shape distortion. This response of SPF specimens to high temperature is contrary to one depicted in freeze-thaw tests (Table 13). Under freeze-thaw cycling, the SPF showed a significant increase in thickness and almost no changes in the other two directions.

The response under simulated field conditions (note the presence of both moisture and thermal gradients) is evidently different from the response to conditions specified by the ASTM D2126 test. The results shown in Table 13 are preliminary in nature; nevertheless, they illustrate how important the selection of test conditions in assessing the dimensional stability of SPF products under field conditions.

5. RATE OF BA EMISSION FROM SPF

At the end of the 1970s, the United States Environmental Protection Agency responding to growing concern about the effects of chlorofluorocarbons (CFCs) on the Earth's ozone layer, sponsored a study, Rand (1980). A model of cumulative CFC emissions from cellular plastics that was developed during this study dealt with three stages of emission: the manufacture (foaming process), the period of service and the products' disposal.

The mass of CFC used in sprayed polyurethane usually ranged between 12% to 14.5% of the foam weight, which is comparable to laminated board products and somewhat less than that normally used in froth systems (16.5%). The CFC loss during field fabrication of SPF was estimated between 11% and 19% of the total load [i.e., 2% to 2.8%, Rand (1980)]. When handling, storage and other incidental CFC loss are added, a more realistic range of 3% to 4% is obtained (Bomberg and Kumaran, 1989).

The emissions from foam surfaces were estimated by calculating the decay of thermal resistance as described by Norton (1967). The half-life period (the age of the foam when half of the initial BA content has been emitted) was used for comparison between different foams. For instance, the half-life period for 50 mm- (2 inch) thick, unfaced polyurethane foam with a density of 35 kg/m³ (2.2 lb/ft³) was about twelve years.

The understanding of aging processes in the foams has been significantly improved since then. This improved understanding, particularly the inclusion of the BA solubility in the DIPAC model, permits relating the emission rate to the effective diffusion coefficient for the BA in question.

The emission rate was measured (Tsuchiya, 1988; Tsuchiya and Kanabus-Kaminska, 1993) using an airtight stainless steel box, in which two freshly cut 25-mm thick specimens (area of 70 × 70 cm or 28 × 28 in) were placed. The box was equipped with an air inlet and an air outlet on opposite walls, an electric heater and a fan for air mixing. Clean air was supplied at a rate of 138 liters (15 ft³) per hour, which corresponded to 0.2 air change per hour for the box. The temperature of the box was set at 20°C (68°F). Outlet air was periodically sampled, with the first sample taken 1 day after the foam was placed in the box. The air samples were analyzed by a gas chromatograph/mass spectrometer, and all major components were identified and quantified. The rate of emission of a compound was calculated from the quantity measured, the rate of air flow into the box and the surface area of the sample. The initial rate of CFC-11 emission ranged between 2.5×10^{-10} and 5.5×10^{-10} kg/(m² s). DIPAC calculations gave the initial rate of CFC-11 emission at about 3.3×10^{-10} kg/(m² s).

Figure 12 compares emission rates measured by Tsuchiya and Kanabus-

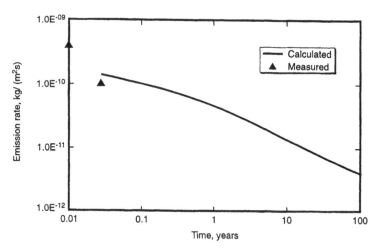

FIGURE 12. Emission rates of CFC-11 measured by Tsuchiya and Kanabus-Kaminska (1993) on SPF and emissions calculated with DIPAC model.

Kaminska (1993) and calculated with DIPAC emission rates for the same SPF product (CFC solubility was 3.2% by weight). Since the comparison of measured and calculated data gives sufficient agreement, one may use the DIPAC model for other calculations. Figure 13 shows CFC-11 emission rates calculated from the DIPAC model. Data measured by Kahlil and Rasmussen (1986, 1989) on sixteen samples of four types of polyurethane foams with average density of 28 kg/m³ (1.7 lb./ft³) are also shown.

Let us calculate the half-life period for a 50-mm-thick SPF, assuming the worst-case scenario (Figure 14). With the 14.5% of CFC-11 per foam weight and 3.5% loss during storage, handling, and foam fabrication, the initial CFC weight is 11% per weight. With SPF thickness of 0.05 m and density of 37 kg/m³, the mass of CFC-11 in one square meter of the foam is equal to: $1 \times 1 \times 0.050 \times 37 \times 0.11 = 0.204$ kg. With two surfaces emitting, the half-life time will be reached when the cumulative emission is equal to 51 gram per one square meter of foam. With the average rate of emission for a twenty year period of 0.9×10^{-11} kg/(m²s), the half-life period will be achieved at $0.051/(0.9 \times 10^{-11}) = 5.7 \times 10^9$ seconds (181 years). This agrees with the experiments of Khalil and Rasmussen (1986, 1989) where half-life time of polyurethane varied between 100 and 320 years.

The CFC diffusion coefficient used in this calculation was 0.15×10^{-13}

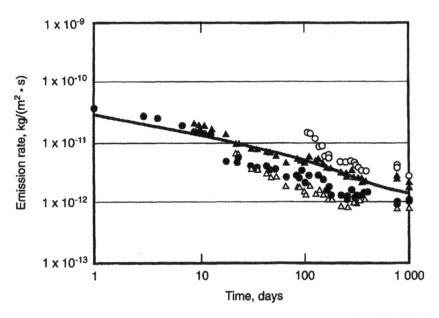

FIGURE 13. Emission rates of CFC-11. Points indicate measurements performed by Khalil and Rasmussen (1986, 1989) on four types of polyurethane foams. The solid line indicates CFC-11 emission rates calculated by using DIPAC model.

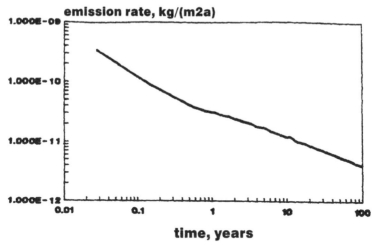

FIGURE 14. Emission rate for 100 years of aging of SPF (with density 37 kg/m³) calculated from the DIPAC model.

m²/s, which is close to the diffusion values reported by Brandreth and Ingersol (1980), but is 10 times lower than that of Norton (1967) or Brehm and Glicksman (1989). The effective diffusion coefficient determined during 1400 days of laboratory aging of HCFC-141b blown SPF was similar, namely 0.3 × 10⁻¹³ m²/s, also indicating a very long half-life period for HCFC-141b-blown SPF.

A good agreement between the measured and calculated emission rates supports the method of extracting the effective diffusion coefficient from thermal measurements (Bomberg and Kumaran, 1989). There is also good agreement between the effective diffusion coefficient obtained from thermal measurements and that based on gas chromatography (Brandreth and Ingersoll, 1980). These results differ, however, from those obtained in accelerated testing[8] (Brehm and Glicksman, 1989).

6. MECHANICAL PERFORMANCE OF SPF

So far, little data has been published on mechanical properties of SPF manufactured with alternative BAs. Yet, without understanding the stress-strain relations, one cannot assess durability and long-term performance of

---

[8]Since solubility changes during the initial period after foam manufacturing are much larger than those at later stage, the accelerated testing may not be as precise as the other measurement techniques.

Table 14. Mechanical properties of MD-type SPF, from Pazia (1995).

Property	Tension	Compression	Sheer
Modulus of elasticity	10.6×10^6 Pa 1553 psi	5.6×10^6 Pa 816 psi	
Strain at yield point		0.063 (6.3%)	
Ultimate strength	385 kPa 56 psi		264 kPa 38 psi
Coefficient of linear expansion	6.2×10^{-2} mm/(m °C) 3.5×10^{-5} in/(in °F)		
Adhesion to concrete block surface	309 kPa 45 psi		

SPF used in different conditions. The mechanical properties of one, MD-type, SPF product, are listed in Table 14.

6.1 Stress–Strain Relation during Compression

One may observe three different regions of the stress–strain curve (see Figure 15).

The first region, the linear elasticity region, is characterized by bending and stretching the struts and membranes of the cells. While some of the membranes may break, most of them remain intact. Typically, this region ends at 6

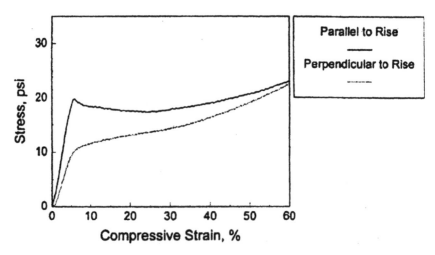

FIGURE 15. A characteristic stress–strain curve obtained under uniaxial compression test. Reprinted from Burns, Singh and Bowers (1998) with permission.

to 8% deformation. Any further deformation will be accompanied by the collapse of the cell membranes.

The second region is called a collapse plateau, since the progress of deformation is met with almost no resistance except for compression of the gas that is still entrapped in the remaining closed cells. This region continues until it reaches such a high degree of compression that struts (polymeric network supporting cell membranes) start touching each other. The next region, cell densification, is therefore characterized by rapid rise of stress with continuing deformation (Gibson and Ashby, 1988).

Since the collapse plateau is wide and easy to measure, Bomberg and Kumaran (1989) used it as a comparative measure of mechanical performance when examining the effect of different BAs on the generic spray foam, Base 88. They also observed the effects of using different polyisocyanate products with the same polyol. The compressive strength changed from 240 kPa to 310 kPa (at 10% deformation[9]) when polyisocyanate was obtained from two manufacturers. The change of BA resulted in a smaller difference (310 kPa and 270 kPa). This fact is not surprising since the reactivity profile and installation technique were controlled and modified to obtain a similar cell morphology for CFC-11 and HCFC-141b foams (Schwartz and Bomberg, 1991).

The above comparison highlights that a change of the BA may not affect SPF mechanical properties if the cell structure and reactivity profiles remain similar. Comparing mechanical properties of CFC-11 blown foams with densities below 64 kg/m^3, and carbon dioxide blown foams with densities above 64 kg/m^3, led to a similar conclusion (Freon, 1985).

Compressive strength measured on different foam products showed large differences. Compressive strength ranging from 100 kPa to 280 kPa (15 psi to 40 psi) may be obtained for 32 kg/m^3 (2 lb/ft^3) foam products. Conversely, foams with densities in the range 24 kg/m^3 (1.5 lb/ft^3) to 40 kg/m^3 (2.5 lb/ft^3) may yield the compressive strength of 200 kPa (30 psi). Furthermore, the compressive strength of SPF depends on the direction of specimen cutting.

The difference between specimens tested in normal (perpendicular to the rise of foam) and foam rise directions may be as large as the difference between various foam products. Even larger differences are obtained when the foam rise is restrained (e.g., molded panels) or when foams are delivered in a pressurized container. Typically, compressive strength for MD-type SPF is 240 kPa (35 psi) in foam rise direction and

---

[9]Although the ASTM test method requires the reporting of strength at the yield point or 10% deformation, whichever comes first, in practice, for determination of compressive strength, only the 10% deformation is used.

Table 15. Average and lowest values of compressive strength of polyurethane foams estimated with 90% confidence interval (Freon, 1985).

Overall Density of the Foam	Compressive Strength	
	Average	Minimum
48 kg/m³ (3 lb/ft³)	410 kPa (60.5 psi)	240 kPa (35 psi)
24 kg/m³ (1.5 lb/ft³)	140 kPa (20.5 psi)	85 kPa (12 psi)

145 kPa (21 psi) in normal direction. Since these differences are closely related to the cell morphology, proper attention must be given to the foam characterization.

For foams with identical cell morphology but different densities, the compressive strength is an exponential function of density. For instance, using the 90% confidence interval and estimating either the average or the lowest compressive strength (Freon, 1985) one obtains the results shown in Table 15.

These results can be compared with data obtained by Kashiwagi and Pandey (1997) on HCFC-141b blown SPF sampled from roof systems (Figure 19). The data, collected in 1995, are not corrected for elevation.

Although different generations of the SPF products are compared, the agreement between Table 15 and Figure 16 is very good. Thus, the relations established on CFC-11 blown foams may still be used, despite a change of the BA.

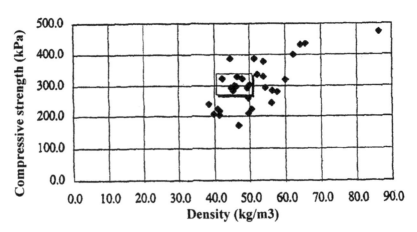

FIGURE 16. Density (overall density) versus compressive strength measured on SPF samples from actual roof installations (Kashiwagi and Pandey, 1997). Reprinted with permission.

Table 16. Average and lowest values of tensile strength of polyurethane foams estimated with 90% confidence interval (Freon, 1985).

Overall Density of the Foam	Tensile Strength	
	Average	Minimum
48 kg/m³ (3 lb/ft³)	550 kPa (80 psi)	290 kPa (42 psi)
24 kg/m³ (1.5 lb/ft³)	250 kPa (36 psi)	140 kPa (20 psi)

6.2 Effect of Density on SPF Mechanical Properties

The agreement between CFC and HCFC blown foam products permits the use of data from an older publication (Freon, 1985). This document reviewed the effects of density on several physical properties of rigid polyurethane foam.

Both tensile and sheer strength have patterns similar to that of compressive strength, i.e., large variations between different foam products and dependence on the cell morphology. These properties can also be expressed as exponential functions of density, if the cell morphology of the foam is identical. The tensile strength data are shown in Table 16 and the sheer strength data are shown in Table 17.

Tables 15 through 17 show that the compressive strength had the strongest dependence on the density, while tensile and sheer showed somewhat smaller effects of density. For 24 kg/m³ (1.5 lb/ft³) density foam, both the compressive and sheer strength values were almost identical.

If change in density had the same effect on the compressive and tensile strength and other mechanical properties, one could use the average or "overall" density when analyzing the mechanical performance of the SPF. However, since this is not the case, one must pay attention to the density variation within the foam.

Furthermore, the stress transfer in polyurethane foams always has three-dimensional character. Cellular structure causes the interaction in other directions, even under a one-directional tension or compression test. One must,

Table 17. Average and lowest values of sheer strength of polyurethane foams estimated with 90% confidence interval (Freon, 1985).

Overall Density of the Foam	Sheer Strength	
	Average	Minimum
48 kg/m³ (3 lb/ft³)	290 kPa (42 psi)	200 kPa (29 psi)
24 kg/m³ (1.5 lb/ft³)	140 kPa (20 psi)	85 kPa (12 psi)

therefore, analyze these locations where the cell-morphology changes. These locations are normally associated with the interface of the "old" and "new" foam, either at the interface between two subsequent passes (lifts) or at the knitline, where one foam pass meets the adjacent pass in the same foam layer.

There is another critical factor affecting the deformation characteristics of the SPF, namely, the nature of the BA. While the change of BA did not affect the overall stress/strain relations, this is not the case with the dimensional stability of the foam.

The role of the BA is highlighted in the published data (Freon, 1985). Comparative testing was performed on polyurethane foam with density of 32 kg/m^3 (2 lb/ft^3), manufactured with two BAs. Increasing the test temperature from 70°C to 100°C (from 158°F to 212°F) caused an increase in the linear, dimensional change from 2% to 6% for the foam blown with carbon dioxide. For the same density CFC-11 blown foam, the corresponding change was from 3% to 18% linear dimensional change. While the changes in the cell-gas pressure are identical, the mechanical properties of both foams are similar and the difference can be associated with the change of the BA.

6.3 Effect of BA on Dimensional Stability of SPF

A change of the BA may affect dimensional stability, in either a "chemical" or a "physical" manner. The effect of "chemical" changes on foam dimensional stability is best seen when high temperature is coupled with a high relative humidity of air, simultaneously with significant physical effects. Such a phenomenon is shown in Figures 17 and 18.

Each of two SPF batches were exposed to two environments: a dry heat environment (70°C and ambient RH) and a hot and humid environment (70°C and ($\Phi > 99\%$ RH). The dimensional change measured after fourteen days for foams exposed to dry heat was about 1%, but those exposed to a hot and humid environment showed 4% and 8% respectively.

Figure 18 shows results obtained on the same generic SPF manufactured with progressively increased amount of water (amount of CO_2). While the product (a), manufactured with 0.5% of water shows behavior identical to that without water, products (b) and (c) manufactured with 1.0% and 1.5% of water, show expansion to 2 or 3%, followed by shrinkage.

Figure 18 implies that membrane breakage occurred during the hot and humid exposure. The broken membranes released some entrapped gas, reducing cell-gas pressure, which resulted in shrinkage of the specimen. Since this cell breakage occurred at elongation much smaller than the yield point of the SPF product, one may assume that the addition of water modified the elasticity of the polymeric matrix. This observation is similar to the one reporting dramatically different dimensional stabilities of

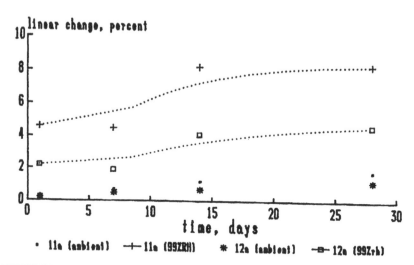

FIGURE 17. Dimensional changes measured for two batches (difference in polyisocyanate) of the generic SPF (Base 88) under exposure to 70°C and two levels of relative humidity—either ambient air or 99% RH. Reprinted with permission of NRCC.

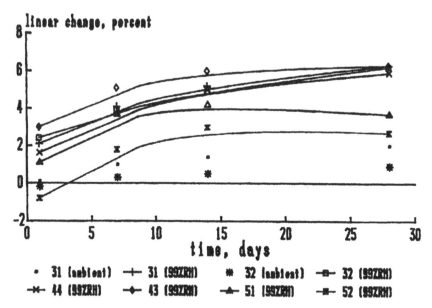

FIGURE 18. Dimensional change under two exposures (70°C and either ambient or 99% RH) of Base 88 manufactured with: (a) 0.5% water, (b) 1.0% water, (c) 1.5% water. Reprinted with permission of NRCC.

polyurethane foams manufactured with CFC-11 and carbon dioxide (Freon, 1985).

Bomberg and Kumaran (1989) measured other characteristics of these three foams, namely water vapor permeance, oxygen, nitrogen and BA diffusion coefficients, and cell morphology. They noted that while small water contents resulted in insignificant changes, a higher percentage of water addition appeared to have reduced uniformity of the polymer matrix.

Figure 19, quoted from the research of Smits and Thoen (1990), highlights the effect of the BA pressure on dimensional changes. The example, derived from computer calculations, shows cell pressure as a function of time for three cell gas mixtures (1) 100% of CFC-11, (2) 50% of CFC-11 and 50% of CO_2, (3) 100% of CO_2. Figure 19(b) shows dimensional changes of the foam filled with 100% of CO_2.

A significant initial shrinkage (about 8% by volume) is observed. This shrinkage is slowly reduced until, after about 26 days, the specimen recovers its original size. One may assume that the process of air ingress is completed. Figure 19 provides clear understanding of the time-dependent effects, namely, how the cell-gas pressure changes affect the overall dimensional stability of the foam.

Furthermore, as the changes in the cell-gas pressure in the center of a thick foam layer are delayed in relation to its surface, there is also a spatial distribution of pressure. Therefore, comparative testing must be performed on specimens with identical dimensions.

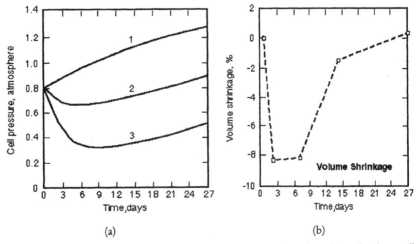

(a)   (b)

FIGURE 19. Cell pressure (fraction of one atmosphere) as a function of time for three cell-gas compositions (1) CFC-11, (2) CFC-11, and CO_2, (3) CO_2 and volume shrinkage of the carbon dioxide blown form. From Smits and Theon (1990), with permission.

In effect, the main factors affecting the cell-gas pressure are:

- boiling point of the BA
- BA solubility
- thickness of the foam (assumed constant for comparative purposes)
- "elasticity" of the polymeric matrix

For HCFC-141b, three of these factors result in increased pressure gradients in the foam. Firstly, the boiling point for HCFC-141b is higher than CFC-11 which may reduce the initial cell-gas pressure, typically from 0.8 atmosphere, as shown in Figure 19 for the CFC-11, to 0.73 atmosphere for HCFC-141b blown foam. Secondly, higher solubility of HCFC-141b will reduce the equilibrium pressure in the center of the foam, typically from 0.45 atmosphere for CFC-11 blown SPF, to 0.39 atmosphere for HCFC-141b blown foam.

Thirdly, the polymer softening (plasticization of polyurethane matrix) gives a small but measurable reduction of the compressive strength of the polyurethane foam. Using results published by Sing et al. (1995), the reduction between 6% and 12% of compressive strength is estimated (for HCFC solubility values shown in Table 9).

Figure 20, based on a computer model (DIPAC, see Bomberg and Kumaran, 1995), shows the cell-gas pressure in the surface layer and in the middle of 50 mm thick SPF as a function of time. After 40 days of aging, the difference in cell-gas pressure between the surface and the center is about 78 kPa for the foam manufactured with CFC-11 and 90 kPa for the foam manufactured with HCFC-141b.

Figure 20 highlights the point that while the surface layers have already reached equilibrium with the atmospheric pressure, a strong shrinkage in the center of the foam prevails during several months of the foam aging. Whether this differential stress can cause an overall shrinkage or foam delaminating at the pass (lift) interface, (interface between two subsequent applications of foam) depends on the mechanical properties of the foam.

The minimum sheer strength, listed in Table 17, was 85 kPa for 24 kg/m³ (1.5 lb/ft³) foam, i.e., less than the 90 kPa calculated from the DIPAC model. This comparison explains why the density limit for the MD-type SPF was set at 37 kg/m³ (2.3 lb/ft³). The minimum sheer strength for the SPF with this density is approximately 150 kPa (22 psi).

Such a high limit for the sheer strength is selected because of possible moisture presence in the foam. For SPF with 1.3 atmosphere total cell-gas pressure (18.5 psi) at 20°C (68°F), when solar radiation increases surface temperature to 70°C (158°F), the pressure of the dry cell-gas pressure will reach 165 kPa. The partial pressure of water vapor is 33 kPa (4.52 psi), which brings the total cell-gas pressure to 198 kPa (29 psi). The center of the foam, however, remains

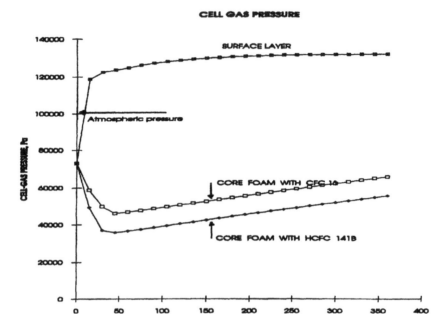

FIGURE 20. Cell-gas pressure in the middle and surface layers of SPF manufactured with CFC-11 and HCFC-141b according to DIPAC model calculations.

at room temperature, and the cell pressure is about 40 kPa. In effect, the differential sheer stress of 158 kPa can be expected, i.e., about the limit of the sheer strength acceptable for the MD-type SPF.

The total cell-gas pressure of moist foam at 70°C (158°F) was about 200 kPa. This corresponded to the tensile strength of 37 kg/m³ (2.3 lb/ft³) foam. In effect, the practical experience agrees well with the calculated stress levels.

Finally, Figure 20 also shows that the traditionally used fourteen or twenty-eight days period of testing for SPF dimensional stability is not sufficient, and other means of accelerated testing must be employed.

6.4 Forces on Pass and Knitline Surfaces

During the foam rise stage, the cell gas exceeds the atmospheric pressure. This causes foam expansion in all three directions. While in the field, the foam is restrained by the adjacent cells and can only grow in the normal direction; this is not the case at the edges. A two- or three-dimensional foam expansion takes place at the edges (this phenomenon is called the creep of the foam).

Furthermore, the cells of the spray foam, which in the field are elongated in a normal direction, may show elongation in the lateral (horizontal) direction at the edges. Such cell orientation usually reduces compressive strength of the foam.

Another phenomenon that occur at the foam surface is a loss the BA which causes a change in cell morphology (surface densification). While designing SPF systems, manufacturers try to reduce the presence of "skins," but their complete elimination is not possible. There will always be some change in cell morphology at the foam surfaces. How much, change, depends on such parameters as age of the chemical in drums and conditions of installation (ambient temperature, pressure and foam temperature, nature of the substrate, etc.). This change has been particularly noticeable when applying SPF to concrete or masonry surfaces.

The surface densification and differences of the elasticity on the interface between "old" and "new" foams do not have practical significance if the next layer is applied correctly[10](see Chapters 6 and 7). Often, in the early stage of SPF use, the foam was applied on a large area of the roof and the second pass application was delayed until the next day. If application of the second pass was delayed too long, the surface of first layer of the foam already equalized with the ambient air. The cell-gas pressure in the first SPF pass is about one atmosphere (or higher), while the cells of the second layer ("overspray") undergoes cooling to ambient temperature. The second layer undergoes shrinkage with cell-gas pressure becoming as low as 0.4 atmosphere. The new foam sets under conditions of strong differential pressure and significant shear stress acting in the pass plane. This factor contributed to some delamination failures in earlier foam installations (Kashiwagi and Moor, 1986).

As the chain breaks at its weakest link, the interface created by the "old" and "new" foams is potentially the "weakest" point of any SPF system. Although the densification of the foam surface alone does not lead to foam damage, it may be one of the factors causing foam delamination when occurring simultaneously. These factors are:

- large differences in the cellular structure of skin and core
- large differences in water vapor permeance of skin and core, (leading to a local accumulation of moisture)
- insufficient blending of old and new SPF layers (variation on the knitline)
- variation in uniformity of the polymeric matrix

---

[10]Joints between old and new pass should not occur at the same location. Installation of the spray foam is usually started in the vicinity of junctions (or places where a shear may occur) and the first, pass is usually somewhat thinner. Some installers oscillate between passes applied with horizontal strokes and those applied with vertical strokes.

- polymer softening caused by presence of moisture
- other effects of moisture entrapment

As previously mentioned, the presence of moisture is a significant factor because of two reasons: polymer softening and its effect on total cell-gas pressure. Assume that condensed water is present in the foam layer adjacent to the upper surface. Assume that the aged foam with a total cell-gas pressure of 132 kPa (18.5 psi) at 20°C (68°F), and its surface temperature increases to 70°C (158°F). Then, the total cell-gas pressure increases to 198 kPa. Even more pronounced is the effect of water in a hot and humid climate. At a roof temperature of 85°C (185°F), the dry cell-gas pressure reaches 172 kPa, but partial pressure of water vapor of 60 kPa (8.4 psi) would bring the total cell-gas pressure to 233 kPa or 34 psi. This corresponds to the limit of the sheer strength of HD-type SPF, see values for 48 kg/m³ (3 lb/ft³) in Table 15.

Again, this strength requirement is confirmed by the roofing practice. Experience indicates that SPF density should be a minimum of 48 kg/m³ (3 lb/ft³) for exposed areas—coastal areas of Florida or California and Alaska, while the minimum density of 40 kg/m³ (2.5 lb/ft³) may be sufficient in other areas.

The calculations of stress in hot and humid environments draw attention to the importance of moisture management in SPF systems and highlight the need for uniformly good adhesion between the coating and SPF layer in roofs.

7. THERMAL PERFORMANCE OF MOIST SPF

This topic is frequently misunderstood. Some researchers believe that laboratory measurements in a Heat Flow Meter apparatus (ASTM C 518) when the heat flux has already been "stabilized" can be used to define field thermal performance of moist insulations.

Actually, in the initial stages of moisture redistribution, the apparent rate of heat transfer may be stable through a prolonged period. Bomberg and Shirtliffe, 1977, reported a test when an apparent and stable for almost two days. Yet, during this initial two day period the measured heat flux differed more than 20 percent from that obtained during the final (dynamic) equilibrium.

To understand thermal performance of the moist material, one may examine the effect of the redistribution of limited amount of moisture added to the insulation surface. Such an experiment was performed by Kumaran (1989), who took a slice of thin fiberboard, added 15 grams of water to it, and placed on the hot side of the specimen, as shown in Figure 21. Using four types of materials,

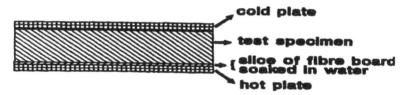

FIGURE 21. Experimental arrangement in demonstration test with four materials (Kumaran, 1989).

defined in Table 18, Kumaran (1989) examined the rate of heat flux and the moisture content profiles.

Figure 22 shows the "anecdotal" and stable levels at the initial stage as being almost three times higher than those measured for the dry specimen of the glass fiber. The cellulose fiber shows initial heat flux double at the dry heat flux, but effect of moisture lasts much longer than that for glass fiber. On the other hand, the extruded polystyrene and SPF specimens show no change in the heat flux during the whole test. The increase of heat flux caused by the moisture transport is small for both materials (this effect was slightly higher in the SPF).

To explain these results one must look simultaneously at heat fluxes and distributions of moisture content across the specimen. The latter were measured with a gamma-spectrometer at different intervals of time counted from the start of the experiment.

Specimen (a), high density glass fiber board, displayed a rapid transfer of water vapor with condensation first occurring at layers four or five (about 9 mm, or 3/8 inch from the hot side). While the first "portion" of water evaporated and diffused further, a new quantity of water vapor condensed. Initially, the quantity of water condensing and evaporating and, therefore, the contribution of latent heat was almost constant for about five hours. This creates a "quasi-stationary regime" in which the heat flux is stable. A less experienced operator would believe that conditions of ASTM C

Table 18. Characteristics of tested materials.
Thermal conductivity measured at 24°C.

Specimen Code	Type of Insulation	Density kg/m³	Thickness (mm)	Dry Thermal Conductivity W/(m K)
a	Glass fiber	63	25.2	0.0311
b	Spray cellulose	45	25.2	0.0405
c	Spray polyurethane	39	25.8	0.0178
d	Extruded PS	30	25.3	0.0256

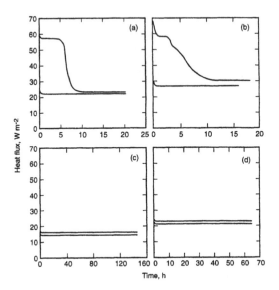

FIGURE 22. Heat flux through a dry (lower curve) and wet specimen (upper curve). Specimens to which 15 g of moisture was added on the hot side are described in Table 13 (Kumaran, 1989). Reprinted with permission of NRCC.

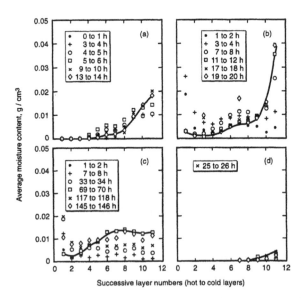

FIGURE 23. Moisture content profiles at different intervals of time counted from the beginning of the test. Specimens are described in Table 13. (Kumaran, 1989). Reprinted with permission of NRCC.

518 test method are satisfied. The "anecdotal" proof that the thermal conductivity of wet material is almost three times lower than that of the dry one is thereby achieved.

The efficiency of the condensation/evaporation cycle is then reduced. Between the ninth and tenth hours of the experiment, one may observe the peak of the moisture content at the cold side. Now the moisture evaporation is constant again, but all moisture evaporates and leaves the specimen. No further effects of moisture on heat flux are measured.

Specimen (b), cellulose fiber insulation, being more hygroscopic, does not show such clear effects of evaporation and condensation. Moisture content of the layer adjacent to the hot side stabilizes during the first half hour and for the next two hours, a similar "quasi-stationary" state occurs. The contribution of condensation/vaporation cycle is then successively reduced. Comparing moisture content distribution after twelve hours with that measured after eighteen hours one may notice that moisture moves from the center to the cold layer. This layer still accumulates more of the moisture passing through it. Overall contribution of evaporation, as shown in Figure 23, however, is reduced.

Specimen (c), SPF and specimen (d), extruded polystyrene, show similar behavior in one respect. Namely, there are no changes in the mass transfer contribution during the whole period of measurements.

The SPF specimen shows that small but constant condensation and evaporation take place almost everywhere within the foam. However, since water vapor permeability is small, latent heat transfer affects the rate of heat flow only to a minor degree.

Specimen (d) shows very little condensation inside the foam, and moisture accumulation takes place only at the layer adjacent to the cold plate.

Since the thickness of these specimens was the same, the level of heat flux shown in Figure 22 indicates the thermal insulating performance.

The above experiment involved small quantities of water delivered to the hot surface of the specimen. How would thermal performance of the foam be affected if moisture was contained inside the specimen and thermal conditions were dramatically changed? Figure 24 shows such an experiment. A specimen was taken from a prolonged soil exposure (the exposure site was protected from direct sun radiation). The specimen was placed in the HFM apparatus. Although stabilization of heat flow took place within one day, the test was continued for almost four days. After ninety hours, the apparatus was opened, the cold and hot surfaces of the specimen were reversed and the test was continued. A 15% increase in the heat flux was observed. This increase decayed rapidly, and within one day, heat flux reaches the same level as before the reversal.

The experiment shown in Figure 24 implies that, under typical service

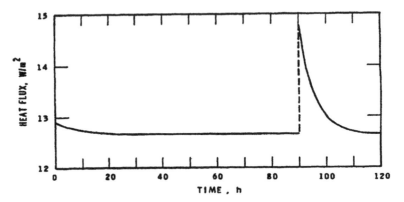

FIGURE 24. Heat flux through a wet SPF specimen taken from soil exposure and placed in a heat flow meter apparatus. The cold and hot surfaces were reversed after 90 hours. Reprinted with permission of NRCC.

conditions, even when using the flow-through design i.e., allowing some accumulation of moisture in the material, thermal performance of SPF products will not be much affected by moisture.

8. OTHER PERFORMANCE CHARACTERISTICS

8.1 Sound Absorption

Table 19 (from Freon, 1985) lists the sound absorption coefficient measured on 32 kg/m³ (2 lb/ft³) polyurethane foam for sound frequencies ranging from 125 to 4000 cycles per second (Hertz).

The closed-cell structure of SPF does not produce good sound absorption. Nevertheless, SPF can contribute to noise reduction by reducing the "drumming" effect of some sheet materials and by reducing air-born spread of noise.

The OCF-type of SPF may have a significant effect on sound reduction if applied in a sufficient thickness. Table 20, (from Ball et al., 1960), lists the sound ab-

Table 19. Sound absorption coefficient of
polyurethane foam with density of 32 kg/m³ (2 lb/ft³).

Frequency Cycles/s	125	250	350	500	1000	2000	4000	NRC
Absorption coefficient	0.12	0.18	0.20	0.27	0.19	0.62	0.22	0.32

Table 20. Sound absorption coefficient of
flexible foams with density of 34 kg/m³ (2.1 lb/ft³).

Frequency Cycles 1/s	Thickness (mm)	12	250	350	500	1000	2000	4000	NRC
Absorption coefficient	12	0.1	0.21	0.21	0.26	0.17	0.63	0.65	0.31
Absorption coefficient	25	0.1	0.25	0.29	0.45	0.84	0.97	0.87	0.63
Absorption coefficient	50	0.2	0.49	0.62	0.81	0.91	0.98	0.97	0.80

sorption coefficient measured on 34 kg/m³ (2.1 lb/ft³) flexible urethane foam for the sound frequencies ranging from 125 to 4000 cycles per second (Hertz).

If a thin, impermeable surface film is applied over the flexible foam, it may alter the impedance of the system. Then, even 12 mm (1/2 in) sprayed flexible foam with 1 to 3 mils film may provide good sound absorption quality (Ball et al., 1960).

In short, OCF-type SPF can be used as acoustic sealing in party walls. For example, STC 59 was measured on a wall with 5/8" drywall on both sides, double studs, double plate, glass fiber blanket R-11, nominal 1-1/2" air space that was sealed with OCF-SPF. Measurements were performed in accordance with ASTM E 336.[11]

8.2 Other Thermal Properties

The specific heat of the SPF is reported (Tye, 1987) as 1590 J/(kg K) or 0.38 Btu/(lb °F).

The coefficient of thermal expansion varies between 3×10^{-5} and 4×10^{-4} 1/K or 1.5×10^{-5} and 2×10^{-4} 1/°F. The lower end of the range relates to closed-cell SPF while the higher values were reported for open-cell foams.

8.3 Vermin and Fungus Resistance

Freon (1985) states "Rats will nibble on rigid urethane foam but will not consume it even if subjected to starvation conditions."

The same source stated that both polyester- and polyether-type foams normally will not support fungus growth. A test with military specification MIL-F-13927 and ASTM D 684 soil burial showed no growth on the specimens. However, since SPF does not contain fungicides, if a fungus colony is transplanted from another site and the right thermal and moisture

---

[11]John Hillard, Santa Ana, CA, private communication.

conditions prevail on the foam surface, the continuing growth of the fungus is possible.

8.4 Prolonged Exposure to Soils

Ten-year burial tests were conducted in Delaware. Tested specimens included pine boards, sections of aluminum-faced panels and specimens of rigid polyurethane (PUR). The panels were buried to 1/2 of their length and PUR specimens had 250 mm (10 in) placed in the soil. The soil exposure lasted for a period of ten years, and specimens were tested after one, three and ten years.

The results obtained on untreated foam panels were excellent. There was a small reduction of tensile strength, an increase in compressive strength, and a slight increase in weight loss in one panel exposed to 121°C (250°F). The results obtained on foam containing flame retardant were quite different. Foam was saturated with water and started to deteriorate after three years. This comparison highlights that the fire retardant effect on both moisture accumulation and on SPF dimensional stability must be recognized.

Pine boards were attacked by termites in three years and rotted in ten years. In contrast, "there were only few holes in the edges of the foam. The PUR foam with flame retardant chemicals had fungus growth throughout the buried portion of the panel."

PUR foams had only a few pockets made by rodents or insects. Except for small increases in volume and accelerated thermal drift (aging) of the samples, there were no substantial differences between the samples aged indoors and those aged in the soil.

8.5 Prolonged Outdoor Exposure

Samples of the sizes similar to those used in the soil burial test but protected with three types of synthetic vapor barriers (one Hypalon® and two Insul-Mastic® products) were exposed to solar radiation on an outdoor rack. Unprotected specimens of polyurethane were also kept indoors for comparative purposes.

The surface of the unprotected, outdoor-exposed foam was badly eroded, while the protected foams were generally in satisfactory condition. Except for some cracking and flaking of the coating, birds pecking at the edge, and partial delaminating from the substrate on the sample, the physical properties measured on the exposed samples were close to the indoor-kept reference sample.

One general observation made when reviewing several tables of test results (Freon, 1985) was that a scatter between measurements performed at different times on control samples was often much higher than the probable effects of soil burial or solar radiation. For instance, in the last exposure tests, the average of the original WVT measurements for all samples was 7.7 ng/(Pa·s·m), i.e., 5.3 perm per inch. If denoting it as 100%, an average WVT after one year is equal to 145%, after 3 years is equal to 83% and after 10 years is equal to 58% of the original WVT. The difference between the reference and other samples after 10 years is 0% for one set of 2 samples, and 10% for the other set of 2 samples.

The compressive strength of laboratory-stored samples showed time-dependent changes. The first year of aging brought a significant increase in compressive strength. This increase was much smaller in the next two years and was followed by a small reduction during later years. In effect, after ten years, these samples showed approximately the same level of compressive strength as those measured after one year. After ten-year exposure, two of the foam samples showed additional reductions, bringing the compressive strength to the level of the initial measurements. Two other samples showed the same compressive strength as the foam stored in the laboratory.

CHAPTER 3

Testing Physical Characteristics of Foams

THIS SECTION REVIEWS different laboratory procedures used for character-izing SPF products and determining their physical properties. These test methods are repeatable and reproducible, however, they may or may not be correlated with the SPF field performance.

1. CHARACTERIZATION OF FOAM STRUCTURE

Normally, a statistically significant number of specimens must be tested to characterize performance of the commercial product. This number may be reduced if one can characterize cellular structure of the foam, show that specimens prepared from different product batches are similar, and use the existing database (Bomberg and Brandreth, 1990). Some measurements used for characterizing foam structure and quality are discussed below.

1.1 Preparation of SPF Sample for Laboratory Testing

Unless otherwise specified, 12 mm- (1/2 in.) thick plywood or 14 mm (5/8 in.) waferboard sheets are used for the standard sample preparation. It may be convenient to spray on a [0.9 m × 0.9 m- (3 ft × 3 ft)] sheet and cut this sheet into smaller specimens just before performing the actual tests. During the foam application, ambient temperatures should be close to room conditions, namely 24 ± 4°C (74 ± 6°F), and the relative humidity should be in the range of 30% to 80% RH.

Spray about 25 mm (1 in.) foam layer on the central 1 m (3-1/2 ft) part of the sheet and rotate the sheet.[12] Then, apply either an approximately 60 mm- (2-1/2 in.) thick layer for tests of HD-type or an approximately 100 mm (4 in.) thick SPF layer. The thickness of a pass should not be less than 12 mm (1/2 in.) or more than 40 mm (1-1/2 in.) for the HD-type SPF and no more than 30 mm (1-1/4 in.) for the MD-type SPF. Normally SPF

---

[12]Application on both sides of the sheet prevents buckling of the layer during the curing period.

passes are applied with three to five minute intervals between them. Alternatively, three passes of the MD-type SPF are applied to achieve the required thickness.

1.2 Core Density

Density is determined on the core specimens, obtained by removing 3 to 6 mm (1/8 to 1/4 in.) trim on both external surfaces (ASTM test method D 1622). Core specimens may contain one or more internal skins at pass interfaces (ASTM specification C1029). Normally, three 50 mm- (2 in.) thick, 600 mm × 600 mm- (24 in. × 24 in.) square specimens are used for determination of the average density.

1.3 Characterization of Cell Morphology with Image Analysis

One of the most accessible methods of foam characterization creates a digital image of the cellular network. This image is created by measuring the light transmission through a thin slice of the foam. The Image Analysis (IA) method was examined with a view to determine cell-size distribution, cell orientation (aspect ratio), and distribution of polymeric material in the struts.

Schwartz and Bomberg (1991) proposed characterization of cell morphology which includes mean cell diameter (d), mean volume-surface diameter (d_3), aspect ratios (A_r, A_n), and an estimate of the minimum fraction of continuous struts (f_1). A sample preparation technique was also described in the discussed paper.

Some of these features were found to be precise enough to be used for comparative purposes within one laboratory (with many operators). These features were cell diameters (d) and (d_3) and aspect ratios in rise (A_r) and normal (A_n) directions. Other features, especially those based on a thinning technique, were not suitable for comparisons involving more than one operator.

Table 21 lists three features: d_1 is the mean diameter of windows, d_3 is the mean volume-surface diameter of windows, and A_r is the aspect ratio. The mean cell size is then estimated using the mean d_3 diameter.

Table 21 shows IA measurements carried out on a few sprayed polyurethane foams manufactured with HCFC-141b. The SPF products showed elongation in the rise direction. The degree of elongation was different, from a moderate elongation of products S and M, to a significant elongation of products P1, K , and R. Of particular interest are two products, denoted as P1 and P2, manufactured with identical polymer but using CFC-11 and HCFC-141b respectively. This foam was optimized for the

Table 21. IA characteristics measured on SPF products.

Code	Area μm^2	Cell Diameter*		A_n	f_1-Factor
		$d, \mu m$	μm		
P1 normal	1650	112	274	1.00	0.15
P1 rise	3418	162	370	1.55	
P2 normal	1901	121	290	1.09	0.16
P2 rise	3017	152	365	1.37	
R normal	1114	92	221	1.09	0.22
R rise	2671	143	390	1.52	
S normal	1463	106	254	1.03	0.14
S rise	2230	131	321	1.21	
K normal	—	208	510	1.05	—
M normal	—	194	475	1.01	—

*Cell size was calculated using an average window area.

use of HCFC-141b, and the cell structure appears to be very similar to the CFC-11 blown foam.

1.4 Foam Characterization with Normalized Aging Curves

The measurement of the oxygen or nitrogen diffusion coefficient may also be used to characterize the SPF. It has been postulated that the effective diffusion coefficient of a simple gas would describe the transport of gas through the pinholes, ultra-thin membranes, and other imperfections in addition to that through "typical" membranes. Thus, comparing diffusion of the same gas through the film and the foam slice, one could assess the degree of foam imperfections.

While the effective diffusion measurements are normally performed separately for each gas, their combined effect can also be used for the foam characterization. Such a case, involving all transport processes (diffusion of oxygen, nitrogen, BA, and solubility of BA that are occurring simultaneously), is represented by the normalized aging curve.

The normalized thermal resistivity (relation between the actual thermal resistivity and its initial value) has been shown as almost independent of the initial cell-gas composition and strongly dependent on the polymer and cell morphology (Baumann, 1982; SPI 1988). This is illustrated by Figure 25, which shows four aging curves (thermal resistivity versus aging period), for the same SPF applied on different substrates. Figure 26 shows the same data with normalization applied.

Even though the initial thermal resistivity of SPF surface layers differed by as much as 20%, the normalized aging curves are alike, indicating similar diffusion rates for all four tested slices.

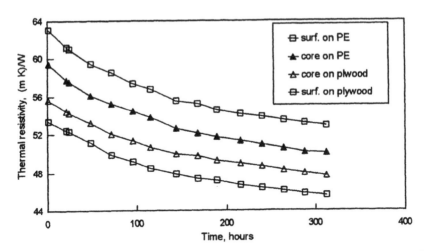

FIGURE 25. Aging curves (thermal resistivity in relation to its initial value as a function of time) for the same polyurethane foam sprayed on two different substrata. Reprinted with permission of NRCC.

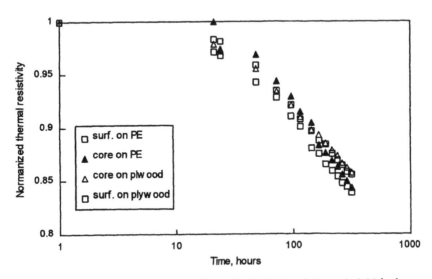

FIGURE 26. Normalized aging curves (thermal resistivity in relation to its initial value as a function of time) for the same polyurethane foam sprayed on two different substrata. Reprinted with permission of NRCC.

1.5 Maximum Temperature and Time of Its Occurrence (Reactivity)

Spray an approximately 40 mm- (1 1/2 in.) thick layer of SPF on the selected substrate. Immediately after spraying the first SPF pass, place a temperature sensor (e.g., thermocouple) on the foam surface, fasten it with a piece of tape or a non-metallic clip, and record time (e.g., press a stopwatch). Apply the second 40 mm- (1 1/2 in.) thick layer of SPF as soon as possible.

Periodically, record both time and temperature to extract the shortest period of time at which the maximum temperature was achieved. This test is used for the initial characterization of the SPF system as well as for the quality assurance in the SPF fabrication process.

1.6 Compressive Strength of the Green Foam

Determination of compressive strength on green (freshly made) foam is necessary for characterization of SPF. The curing period required before the test, is minimum two hours in room temperature. If the test is not performed within 24 hours after the foam fabrication, the report should include foam age in days.

The ASTM standard test method requires the measurement of the compressive resistance (strength) at the yield point or at 10%, whichever comes first. The yield point, which normally occurs at smaller deformation, however, may give a small bump on the stress/strain relation. The practice within the industry, therefore, is to determine the compressive strength at 10% deformation. This is of no practical consequence since the increase in cross-linking density that takes place with time may cause a subsequent increase in the compressive strength as high as 30%–50%.

1.7 Other Means of SPF Characterization

In addition to density, thermal properties, and dimensional stability, one may need to characterize the mechanical performance of the SPF. Measuring compressive strength[13] with the load applied parallel to the foam rise direction is commonly used for this purpose.

The standard specification ASTM C1029 lists other foam characterization tests: the closed-cell content, water vapor permeability, water absorption, and tensile strength. We do not recommend using these tests for foam characterization for the following reasons.

---

[13]Compressive resistance, ASTM C165, procedure A, tested with the head speed 2.5 mm/min.

- To measure the closed-cell content, one must use both procedures A and B as specified by the ASTM D2856 standard, and average their results. Such measurements are time consuming and the judgment value of this test is poor.
- Water vapor permeability and tensile strength are important characteristics, but they depend on the polymeric composition as much as the distribution of foam imperfections in the polymeric matrix. In effect, these tests require a large number of measurements.
- The ASTM D2842 test method is applicable for measuring water absorption of foams used for buoyancy applications. Much better insight into moisture performance of the foam can be obtained using water absorption under thermal gradient conditions (see later text).

Another method of foam characterization during the field fabrication involves spraying a test pattern.

2. LONG-TERM THERMAL RESISTANCE OF FOAMS

SPF, like many other foam insulations, is subject to aging (thermal drift) processes. Aging means that thermal resistance of the material changes during its service period, mainly because of the changes in composition of the gas contained within the closed cells of the foam. Since aging is a very slow process and occurs over many years of service, one can reduce the period of laboratory testing, for instance, by measuring the aging of thin layers. However, aging of small specimens is rapid and can only be applied if the initial thermal performance of the actual foam product is determined using large specimens. Either a hot box method (ASTM C-236) or a HFM apparatus using 600 mm × 600 mm-specimens was found appropriate for determination of the initial R-value.

The following steps are necessary to estimate long-term thermal resistance (LTTR) of the SPF product:

1. Determine a mean value of the initial thermal resistivity of the product.
2. Measure initial thermal resistivity prior to slicing the specimens.
3. Cut thin layers, keep in the laboratory and periodically measure their thermal resistance.
4. Determine the aging factor as the ratio of thermal resistance at the end of predetermined aging period to the initial thermal resistivity.
5. Determine the LTTR by multiplying the aging factor and the mean initial thermal resistance of the product.

2.1 Initial Thermal Resistance of the Product

To establish the initial thermal resistance of the SPF product, thermal resistance tests must be performed on five to seven, specimens, and their results must be averaged. Determination of the initial thermal resistivity normally involves testing specimens cut from one to two week-old, 75–100 mm- (3–4 inch) thick slabs. As an absolute minimum, three large specimens can be used to establish the initial thermal performance of the SPF product (see Appendix A1).

2.2 Initial Thermal Resistivity Measured on Thin Layers

To study the aging process, thermal resistance of thin layers is measured at the prescribed stage of aging.[14] A recommended procedure for LTTR determination is described in Appendix A1.

2.3 Thermal Resistivity of Slices Aged in Room Conditions

To use the ASTM C 1303 procedure, one must select an aging period for which thermal resistance is being evaluated. Should it be the service life of a construction element? Not necessarily, as the expected service life of thermal insulation varies. Typical service life is between fifteen and twenty years for roofs, but period of forty to fifty years is usual for walls. Yet, under typical use conditions, thermal insulation will result in such energy savings that the actual cost of the insulation is compensated much before ending its service life. One may argue that fourteen to eighteen years payback is a typical period for construction. Therefore, the 20-year period recommended by Smith (1993) for the evaluation of roofs, or 15-year period was proposed by Canadian Industry through TISSQ[15] as a basis for evaluation of LTTR of cellular plastics.

A period of time in which the required degree of aging is achieved depends on the foam thickness. The relation between the aging periods of foam layers with different thickness is defined by the concept of scaling factors. To reach the same stage of aging the ratio of the square root of time and layer thickness must be constant. If thermal resistance of a 50 mm-thick foam layer is to be determined after three-year period (1095 days), and a 10 mm-thick layer is being aged, the test should be done after 44 days. (The power of the ratio 50/10 is equal to 25, therefore 1095/25 = 44 days). Similarly, for a 7 mm-thick layer the test would be done after 21 days. If a 5 year-period (1825 days) is used for estimating thermal resistance of 50 mm-thick foam layer, the testing period for 10 mm-thick layer is

---

[14]A procedure for estimating the long-term change in the thermal resistance of unfaced closed-cell plastic foams under controlled laboratory conditions is described in ASTM C 1303 standard.
[15]Thermal Insulation Systems Standards and Quality Consortium.

72 days. One may observe that ISO standard[16] addresses even a longer period, it namely specifies testing 10 mm- thick layers after 90 days aging.

2.4 Aging Factor for the Selected Reference Period

The aging factor is the ratio of thermal resistivity at a prescribed stage of aging to its initial value. As discussed in Appendix A1 aging factors for core and surface layers are allowed to differ as much as 12 percent.

2.5 Examples of Measured LTTR

Table 22 shows the initial thermal resistivity of a HD-type SPF manufactured with HCFC-141b.

Four slices (425-162) with thicknesses of about 10 mm were prepared and aged in a laboratory room. The ratio of thermal resistivity measured after 72 days of aging to the initial value of the foam specimen was 0.79. The LTTR of this foam, calculated as a product of the mean initial thermal resistivity of the product and the aging factor, is equal to $0.79 \times 50.5 = 39.9$ (m K)/W or 5.8 (ft² hr °F)/(Btu in.). The example in Table 23 shows that initial thermal resistivity of CFC-11 blown, MD-type SPF is significantly higher than the initial r-value for HCFC-141b blown foam (Table 23).

Four, 10 mm-thick slices were tested. The ratio of the thermal resistivity measured on 72 day-old specimen to the initial value is 0.71 and the LTTR of this foam is $0.71 \times 60.3 = 42.8$ (m K)/W or 6.2 (ft² hr °F)/(Btu in.).

Observe that this foam was actually installed in the frame wall with a mean thickness of 75.8 mm. Applying the scaling factor $(75.8/10)^2 = 57.5$ to calculate the reference time, one obtains $1825/57.5 = 31.8$ days. For a 32-day aging period (Bomberg, 1993), the aging factor is 0.745. This value multiplied by the mean initial thermal resistivity of 60.3 (m K)/W yields the long-term thermal resistivity of 44.8 (m K)/W or 6.45 (ft² hr °F)/(Btu in.).

Another SPF foam manufactured with HCFC-141b is shown in Tables 24 and 25. Four, 7 mm-thick slices (426-116) were tested. After 35 days, the ratio of thermal resistivity to the initial value is 0.767 and the long-term thermal resistivity of this foam is calculated as $0.767 \times 53.5 = 41.0$ (m K)/W or 5.9 (ft² hr °F)/(Btu in.). Four, 10 mm-thick slices (426-199) were tested. After 72 days, the ratio of measured thermal resistivity to the initial value (normalized thermal resistivity) is 0.79. Thus, the long-term thermal resistivity of this foam is calculated as $0.79 \times 49.1 = 38.9$ (m K)/W or 5.6 (ft² hr °F)/(Btu in.).

---

[16]Determination of the long-term thermal resistance of closed-cell cellular plastic thermal insulation, a draft of ISO/TC 163/SC1/WG7 has already passed ISO Committee ballot and is in the stage of final approval.

Table 22. Initial thermal resistivity of sprayed polyurethane product M (HCFC-141b).

Batch Code	Density		Thermal Resistivity	
	kg/m³	lb/ft³	(m K)/W	ft² hr °F/Btu in.
425-129	55.3	3.4	50.4	7.27
425-135	54.6	3.4	50.3	7.25
425-141	55.0	3.4	50.9	7.34
Average	55.0	3.4	50.5	7.26

Table 23. Initial thermal resistivity of sprayed polyurethane foam product C (CFC-11).

Batch Code	Density		Thermal Resistivity	
	kg/m³	lb/ft³	(m K)/W	ft² hr °F/Btu in.
396-50	37.5	2.3	59.9	8.64
396-54	35.0	2.2	59.9	8.64
396-55	35.0	2.2	61.1	8.81
Average	35.8	2.23	60.3	8.70

Table 24. Initial thermal resistivity of sprayed polyurethane product R (HCFC-141b).

Batch Code	Density		Thermal Resistivity	
	kg/m³	lb/ft³	(m K)/W	Imperial
426-101	37.0	2.3	52.6	7.59
426-107	38.7	2.4	52.9	7.64
426-117	34.2	2.1	55.0	7.94
Average	36.7	2.3	53.5	7.71

Table 25. Initial thermal resistivity of sprayed polyurethane product K (HCFC-141b).

Batch Code	Density		Thermal Resistivity	
	kg/m³	lb/ft³	(m K)/W	ft² hr °F/Btu in.
426-170	39.0	2.4	49.5	7.14
426-176	38.3	2.4	49.3	7.11
426-188	31.7	2.0	48.6	7.01
Average	34.0	2.1	49.1	7.09

Table 26. LTTR predicted for 50, 75, and 100-mm thick MD-type SPF products B and C [manufactured with CFC and tested in 5 year wall exposure (see Chapter, 10, Section 4.4)]

	50 mm (2 in)	75 mm (3 in)	100 mm (4 in)
SPF layer thicknesses and aging factor	0.71	0.745	0.775
Thermal resistivity (m K)/W	42.8	44.8	46.6
Thermal resistivity, (ft² hr°F)/(BTU in)	6.20	6.45	6.72

2.6 Discussion of These LTTR Results

Using LTTR as a benchmark to compare thermal performance of foams permits drawing some important conclusions.

Firstly, one may compare initial thermal resistivity of CFC-11 blown spray foam product C (Table 23) with that of HCFC-141b blown foams products (Tables 22, 24, 25). The HCFC blown foams have 12% to 18% lower initial thermal resistance than the CFC-11 blown foam. Yet, the difference between the LTTR values of these sprayed foam products is much smaller. The highest LTTR, shown by product R (Table 24) was only 4% lower than the CFC-11 benchmark. The lowest LTTR, shown by product K (Table 24) was 9% lower than the benchmark foam. Thus, better foam optimization accounts for the 5% difference in LTTR.

Secondly, one may examine the effect of density on the LTTR. Intuitively, the HD-type SPF (higher density, therefore, more BA per volume) is thought to have better LTTR. However, the MD-type, product R (Table 24) had the LTTR equal to 41.0 (m K)/W or 5.9 (ft² hr °F)/(BTU in). This is better than HD-type, product M (Table 21). The latter had LTTR equal to 39.9 (m K)/W or 5.8 (ft² hr °F)/(Btu in.). One may observe that the improvement of the cellular structure appears to have more impact on LTTR than differences in the foam density.

The use of LTTR may assist in the development of SPF systems and optimization of their aging process. Therefore, Canadian foam-plastics industry accepted a more realistic period of 15 years as the basis for long-term evaluation instead of 10 years proposed by Polyisocyanurate Manufacturers Association or 25 or 40 years discussed in academic circles. An average value for a 15 year-period of aging is represented by thermal resistance (reference point) after five years of aging. A unified approach to LTTR may also permit the use of the LTTR concept for estimating the actual long-term performance under service conditions. The LTTR methodology, which is proposed in Appendix A1, yields different thermal resistivity values for 50 mm, 75 mm and 100 mm thick layers of SPF. Assume that for 50 mm thick SPF layer the LTTR was 42.8 (m K)/W or 6.2 (ft² hr°F)/(BTU in) and using the aging curve presented in Chapter 10 Section 4.4, one generates LTTR as shown in Table 26.

3. MAXIMUM SHRINKAGE AND EXPANSION OF SPF

As previously discussed the introduction of BAs with higher boiling points and increased solubility affected the period necessary for testing dimensional stability of SPF. The traditional fourteen or twenty eight-day exposure is insufficient to examine the dimensional changes of the foam.

A modified test method was therefore proposed by the industrial task group[17] for determination of dimensional changes. Similar to the ASTM D2126 test method, the goal of this test is neither to predict end-use product performance, nor produce characteristics adequate for engineering or design calculations. This test is developed strictly for comparative purposes to evaluate the response of polymeric matrix to extreme conditions of exposure and, when applicable, to reduce the period of measurements.

Three specimens should be cut from cured foam, have length/width measured in three lines, thickness in five points, and their weight measured with a 0.01 g precision balance. The specimen preparation process should be as fast as possible and by no means longer than four hours.

For testing maximum shrinkage, specimens should be placed vertically (10 mm apart) and kept for 7 days in a vacuum oven. The temperature of the oven should be 30 ± 2°C and air pressure should be reduced to 6 ± 3 mm (0.2 ± 0.1 inch) of mercury. If the vacuum oven is maintained at room temperature, then the exposure period should be extended to ten days.

After conditioning the specimens should be exposed to −40 ± 3°C, ambient humidity and pressure for the period of 28 days and measured after 1, 2, and 7 days. The measurements are performed on a specimen conditioned for 2 ± 1 hour in the room conditions (23 ± 2°C and 50 ± 5% RH). If −40 ± 3°C exposure is not available, one may use 24 ± 3 hour cycling between two environments: −25 ± 3°C and 23 ± 2°C over a one week period with the weekend exposure to −25 ± 3°C conditions.

For testing the maximum expansion, specimens should be prepared in the same manner as those used for the shrinkage test. Immediately after being cut and measured the specimens should be exposed for the period of 28 days to temperature of 70 ± 2°C and condensation conditions, i.e., 98% RH. Interim measurements of the specimen dimensions are made after 1, 2, and 14 days and the final measurement should be made after 28 days. Each specimen is conditioned for 2 ±1 hour in the room conditions (23 ± 2°C and 50 ± 5% RH) prior to determination of weight and dimensions.

---

[17]ICI Polyurethanes Group, 286 Mantau Grove Rd., West Deptford, NJ.

The maximum expansion test is the one specified by ASTM standard 2126. As discussed in Chapter 2: Factors Affecting Performance of the Foam, the stress introduced by this test exposure (hot and humid aging) corresponds to the limits postulated in the SPF classification system. This test may assist in the development of SPF products.

The maximum shrinkage and the maximum expansion tests are developed strictly for comparative purposes, to evaluate the response of polymeric matrix to extreme conditions of exposure. The results obtained in these isothermal tests do not represent field performance (no thermal effects are involved in the exposure) and should not be used for engineering calculations.

4. WATER VAPOR PERMEANCE

Schwartz et al. (1989) investigated water vapor transport with different experimental setups.

One of them, a so-called modified cup method, exposed the specimen to the extreme humidity range. The near saturation conditions, obtained by exposing one side to water (as in the wet cup procedure E96) were applied to one side of the specimen. The near dryness conditions, obtained by using desiccant (as in the dry cup procedure E96) were applied to the other side of the specimen. This research showed that WVT of SPF products measured with the modified cup method was identical to that determined with the ASTM E96 test method (dry cup procedure).

Since the modified cup method does not require a humidity chamber, this method is recommended for quality assurance testing. A schematic of the discussed setup and the used test procedure are discussed below.

This setup utilized a standard 150 mm (6″) PVC pipe cut into two appropriate sections in which groves were machined to accommodate the O-ring. A 6 mm- (1/4″) thick plastic sheet was adhered to the bottom section to provide a tight water container. A 75 mm- (3) deep and 150 mm- (6″) diameter crystallizing dish was placed on a wire support, at a 12 mm (1/2″) distance from the surface of the specimen and of the covering plate. About 300 g of calcium chloride was placed in the crystallizing dish.

The edges of the specimen, with the same diameter as the dish, were sealed and then mounted airtight on the central portion of the setup with the sealing wax. The set-up was then assembled and placed in a chamber maintained at 21.5 ± 0.5°C and a relative humidity of 50 ± 2% RH. The mass gain of the assembly was recorded periodically.

This setup could be modified by replacing the glass container with one having a large number of perforations in the bottom (e.g., a metal mesh) to provide pathways for vapor diffusion. To prevent the desiccant from falling

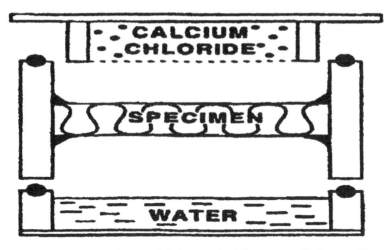

FIGURE 27. Schematic of the modified cup method; ● represents O-rings used for sealing the containers airtight (from Schwartz et al., 1989).

through the perforations one can place a filter paper between the mesh and the desiccant.

5. ABSORPTION AND ACCUMULATION OF MOISTURE IN THE SPF

Three laboratory test methods for determining moisture absorption of thermal insulations were carefully examined and compared in a review (Bomberg, 1983 and Forgues, 1983). Comparison of laboratory and field data (external basement insulation) was also performed. It is generally conceded to be a good way to evaluate the scope of the test method application. This comparison led to the conclusion that moisture gains of the external basement insulations cannot be predicted accurately by any separate test method. Relative assessment is, however, possible from the results of a combination of tests. Furthermore, such a comparison showed how significant differences in moisture absorption are obtained if the same material is exposed to different boundary conditions.

5.1 Water Immersion (Intake) under Isothermal Conditions

The review (Bomberg, 1983) compared six ASTM test methods where specimens are immersed under 25 or 50 mm of water with research performed by different organizations. Direct comparisons were obscured because these test methods used different test conditions. These methods differ in many aspects such as: the period of immersion, the specimen size, the means of removing excess water from the surface (some of them even used underwater weighing), and the allowance or disallowance of interrupted immersion. The review highlighted the significance of the ratio between exposed surface area to the nominal specimen volume and approximation in thickness measurements (see Figure 28).

The review recommended that the water intake test should use specimens placed in contact with water. In scientific studies, a small depth of immersion 2-mm to 3-mm is used and the slope of water intake is usually plotted against the square root of time. In industrial practice, a 10 mm depth was proposed as a practical compromise and the ASTM test method C1134 (1989) was developed to replace all other ASTM test methods except for buoyancy applications of cellular plastics.

5.2 Moisture Accumulation under Constant Thermal Gradient

Thermal moisture drive (TMD) has been identified as a most important mechanism contributing to moisture accumulation in the polyurethane and

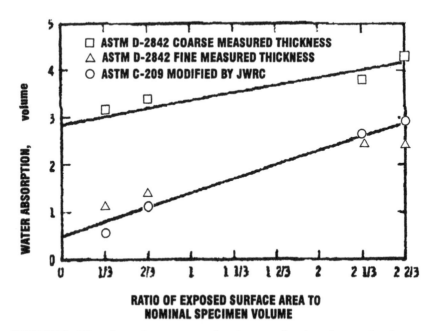

FIGURE 28. Water absorption, percentage by volume, as a function of exposed surface area to the nominal specimen volume. (Data of H. Reid and M. Sherman, see: Bomberg, 1983).

polyisocyanurate foams. A number of different experimental arrangements were reported (Thorsen, 1973; Hedlin, 1977; Dechow and Epstein, 1978; Tobiasson and Ricard, 1979; Norwegian Institute, 1982) and some were even standardized. The European proposal for a standard, "Determination of long-term water absorption by diffusion" (TC 88) builds on the experience of the Swiss Road Institute. The specimen is exposed to 50 ± 1°C on one side and 1 ± 0.5°C on the other side for 28 days. The weight increase is measured once a week. The cold and hot sides of the specimen are then turned into the opposite for the next week exposure. Some research organizations (e.g., Norwegian Road Institute) claim a good correlation with moisture accumulation in road embankments.

In cold climates, one direction of heat and moisture flow prevails over the winter period, while this conditioning procedure simulates moisture ingress from both sides of the sample. This procedure is therefore not recommended for evaluating thermal insulations used to reduce heat transfer.

The test method used to measure accumulation of moisture in SPF measures the accumulation of moisture resulting from the simultaneous exposure to the gradients of water vapor concentration and temperature, see the next section.

6. EFFECTS OF THERMAL AND HYGRIC STRESS ON DURABILITY

Theoretically, exposing a porous material to conditions of one-sided, one-directional environmental cycling that involves freeze-thaw can increase its moisture content until a "failure" occurs (at a specific number of cycles). Such a test would require, however, a much too long period of exposure. After all, most freeze-thaw damages happen after many years of exposure. Conditioning under simultaneous action of water vapor pressure and temperature gradients may cause accumulation of moisture and thereby reduce the number of required freeze-thaw cycles.

Theoretically, even one freeze-thaw cycle may cause damage if the material contains enough moisture (material reached a critical moisture content with regard to freeze-thaw durability[18]). However, such a damage may not happen if the elastic deformation of the material accommodates local stresses caused by freeze-thaw phenomena. Furthermore, if moisture can pass through the material, the material may not reach the critical moisture content under the actual climate and service conditions.

A test method is being developed by an ASTM task group to evaluate du-

---

[18]Critical moisture content with regard to freeze-thaw is usually a high degree of water saturation.

rability of thermal insulation materials under cycling environmental conditions. This method determines compressive resistance of the material (as well as the change in moisture content) in preconditioned specimens placed for 20 days (40 cycles)[19] between two environments:

1. Warm side $T = 24 \pm 2°C$ and $90 \pm 5\%$ RH
2. Twelve hours cycling period of freeze-thaw oscillating between $-15 \pm 3°C$ ($5 \pm 5°F$) and $15 \pm 3°C$, ($60 \pm 5°F$) ambient relative humidity on the cold side

Compressive resistance[20] is measured on the specimens exposed to standardized preconditioning. The strength of the material exposed to the freeze-thaw cycling is then compared with that measured on undisturbed specimens of the same age and taken from the same material batch.

There is also another use of this test procedure. A part of this test (the preconditioning procedure) may be used for benchmarking the computer models.

As previously mentioned, the method of moisture design for building envelopes including SPF should be based on the flow-through principles. Usually, this method requires using a computer model for calculation of the yearly moisture balance. To use the computer model, in turn, one must check if the material characteristics selected for the calculation are correct. This is done during a benchmarking, laboratory test. The preconditioning procedure has been developed in such a manner that it offers an excellent benchmarking opportunity for moisture models.

---

[19]The required number of cycles is always controversial. ASTM C666 standard for testing concrete requires 300 cycles. ASTM E6.58 subcommittee postulated 60 cycles for testing EIFS. Experience with insulating concrete and other porous materials (Chapter 26 of ASTM Moisture Manual) indicates that the number of cycles is not as important as reaching the critical moisture content for a part of the system. Therefore, a stringent preconditioning procedure was mandated by this practice.

[20]Compressive resistance is defined in the ASTM C165 Standard "Test method for measuring compressive properties of thermal insulations."

CHAPTER 4

Evaluation of Protective
Coverings and Barriers

1. INTRODUCTION

WHEN SPF IS applied as an integral part of the roofing system, it must be protected against UV radiation. Typically, the required protection is attained through application of an elastomeric, liquid applied, coating system, which will cure to form a water-resistant protective membrane. As, in principle, the SPF surface does not need any moisture protection treatment, an aggregate layer, supported by a tack coat of asphalt, can also function as the UV protection for flat roofs in many climatic conditions. The selection of gravel or slag is usually determined by local availability; the size being more important than type[21] of the aggregate.

1.1 Design Considerations

A number of factors affect the process of the selection of a suitable protective coating.

Environmental conditions: High tensile strength or abrasion resistance will be required in areas where hail or abrasive conditions are expected. If a chemical attack or pollutants are expected, the coating's resistance to these contaminants must be assessed prior to use.

Flammability requirements: Any system specified should meet all local code and insurance requirements. Independent documentation of ratings should be provided by the appropriate agencies.

Mechanical damage and foot traffic: A coating must be able to resist anticipated mechanical damage and foot traffic. Key physical properties for a coating system to inhibit mechanical damage are tensile strength, elongation, hardness, and minimum dry-film-thickness (DFT). Damage resulting from punctures and other surface stresses can be reduced by using high tensile strength and high elongation coatings. Special

---

[21]See publication SPI Stock Number AY 110.

reinforcements in walkway areas are also used as are rigid panels, pavements, etc.

Water vapor transmission: Selection of protective coatings, in conjunction with spray-applied polyurethane foam, can reduce the likelihood of moisture accumulation within the foam. Install building materials, including foam and coatings, in such a manner that their water vapor permeability (inverse of resistance to water vapor flow) decreases when moving from the side with the higher absolute humidity (usually the warm side). A gradual reduction of WVT characteristics reduces the amount of moisture in the construction.

Ponding water: A roof is "ponded" with water if 24 hours after a rainfall an area of 9 m² (100 square feet)[22] or more holds in excess of 12 mm (1/2 inch) of water. Ponding water should be avoided, whenever practical.

Maintenance: Recommendations for temporary repair and preventive maintenance procedures for use with a coating system should be provided by the manufacturer.

Aesthetics: Many elastomeric coatings can be colored or tinted various shades to comply with job requirements. Using colored granules may also create a uniform appearance.

Other factors: Once the selection process has narrowed to a particular type of coating or the systems of a particular manufacturer, the specifier[23] should weigh these factors:

- field experience
- manufacturer quality control
- warranty
- applicator experience

1.2 Theoretical and Actual Film Coverage

The theoretical coverage is the area covered by a unit of volume of a coating material spread over a flat smooth surface area at a unit of thickness (1 millimeter or 1/1000 of an inch, i.e., 1 mil). However, if the solid content by volume (SCV)[24] is less than 100%, the DFT will be less than 1 mm. For example,[25] a coating with a 60% (0.60) solid content by volume (SCV) will give a thickness of 0.6 mm.

---

[22]A roofing square is a measure equal to 100 square feet.
[23]The architect or a technical assistant who writes technical specifications used for selecting commercial products.
[24]These calculations use solids content by volume, not solids content by weight.
[25]For examples of calculations and conversions between IP and SI units, consult *Spray Polyurethane Foam Estimating Reference Guide,* SPFD/SPI Stock AY 121.

Many factors, such as the surface texture, application loss (overspray), container residue, equipment characteristics, and applicator technique will affect the amount of coating material required to meet the in-place, minimum DFT. Therefore, additional material should be added to the theoretical quantities to ensure that the proper minimum coating thickness is applied (see an industry guide AY-102).

The surface texture is one of the main considerations. Smoother surfaces require less material than rougher surfaces. An excessively rough surface can not be coated since the coating will not provide continuous film without voids, pinholes, etc. Furthermore, mechanically planned surfaces need cleaning vacuum treatment before coating.

Smooth surface: The SPF surface is ideal for receiving a protective coating. Despite classifying this surface texture as smooth, this surface requires at least 5% more material than the theoretical amount. Note that the photographs in Appendix 3 show texture patterns established as the industry reference standard.

1. Orange peel surface—This SPF surface exhibits a fine texture with small nodules and valleys. It is considered acceptable for receiving a protective coating with at least 10% more material than the theoretical amount.
2. Coarse orange peel surface—This surface exhibits a texture where nodules and valleys are approximately the same size and shape. Because of the roundness of the nodules and valleys, this surface is acceptable for receiving a protective coating. This surface requires at least 25% more material than the theoretical amount.
3. Verge of popcorn surface—The verge of popcorn surface is the roughest texture suitable for receiving the protective coating. The surface shows a texture in which nodules are larger than valleys, with the valleys relatively curved. The surface is considered undesirable because it requires at least 50% more material than the theoretical amount.
4. Popcorn surface (also called treebark surface)—This surface exhibits a texture in which valleys form sharp angles. This surface is unacceptable for the coating application, as it is very difficult to completely cover the whole foam surface with the required film thickness.
5. Oversprayed surface—This surface exhibits a coarse textured pattern and/or a pebbled surface. This surface is typically associated with wind affecting the SPF path. The "overspraying" can vary from mild to severe. Mild overspraying requires 25% to 50% more material than the theoretical amount. Severe overspraying is not acceptable for coating applications.

Other factors that affect the actual coverage are:

- *Wind loss:* In spray applications, up to 30% of the coating may be lost due to wind. The operator must consider using wind screens and add wind loss to coating calculations.

- *Miscellaneous loss:* A miscellaneous factor must be added to the theoretical coverage rate to cover losses due to material left in containers, equipment problems, etc. This loss is between 3% and 10%, depending on the contractor's experience and efficiency.

Taking into consideration the minimum DFT, polyurethane foam surface textures, wind, and miscellaneous losses, the total percentage is calculated and added to the theoretical coverage formulas that are published in the SPI guide[26] or by NRCA (1996). Normally, the coating manufacturer specifies both the nominal (average) DFT, and the minimum DFT (in mils), and the operator must consult with the coating manufacturer.

Typically, a coating specification requires a 30 mil coating system (20 mil of the basecoat and 10 mils of the topcoat). This means that the average DFT is 0.76 mm (30 mil) and the minimum DFT is 0.38 mm (15 mil).

1.3 Physical and Performance Characteristics of the Roofing Coating

Physical characteristics that affect the selection of the coating materials are listed below.

- tensile strength and elongation at break
- fire performance of roof coverings
- retention of physical properties upon aging
- water vapor permeance
- adhesion to the polyurethane foam
- UV resistance
- chemical resistance

The first three requirements are the most important. The coating must withstand movements caused by temperature changes. Typically, the coefficient of linear expansion is 6.2 mm/(m °C) (3.5×10^{-5} in./in.°F). If the foam was applied at 20°C and cooled to −35°C during winter (or heated to 75°C in summer), it will expand 3.4 mm/m and shrink 4.7 mm/m. These deformations are well within the yield points of the SPF products, which normally are about 7% in tension and 6% in compression.

Thermal movements of the steel or concrete decks are smaller (coefficient of linear expansion being five or six times lower). Even when correcting for the initial shrinkage of masonry, usually 0.3 to 0.6 mm/m (0.0047 to −0.0074

---

[26]Guide for selection of elastomeric protective coatings over SPF Bulletin AY 102.

in./ft), it is evident that SPF has a much larger tolerance for movement than the structural materials. In turn, tolerance for movements of the elastomeric coatings is another magnitude higher than that of SPF. By definition, the coating must elongate 100% and recover fully (sometimes this is called a 100% memory).

Fire safety is the next main consideration for roof protective coverings. Normally, the SPF and the coating are tested as a composite system under conditions of a fire simulated outside of the roof. Two critical characteristics are examined; namely, the flame spread and the resistance to penetration of fire to the roof deck. The ASTM E 108 standard test method also provides criteria to determine other risks involved in the fire. These risks include an assessment of:

- whether the roof covering material, exposed to a 5.3 m/s (12 mph) wind, would develop flying burning material and, if applicable,
- whether a prolonged exposure to rain would adversely affect the fire performance characteristics of the roofing system.

Most elastomeric coatings used today permit the transfer of water vapor across the coating, yet, in some applications, e.g., cold storage or extreme environments, vapor retarding capabilities may be needed. Table 27 lists water vapor permeance of roof coatings indicating that a designer has a good choice. There are two impermeable coatings, namely, modified asphalts and butyl, two semi-permeable (chlorinated synthetic rubbers), and three permeable coatings, namely, polyurethane elastomers, acrylics, and silicones.

All elastomeric coatings designed for SPF roofs have good adhesion to the polyurethane foam, and medium to high weathering resistance. Therefore, the choice of the coating (or a combination of two coatings) depends primarily on the projected service conditions (slope, traffic loads, aesthetics, budget, climatic conditions).

The following performance characteristics are normally considered:

Table 27. Typical water vapor permeance of generic SPF coating.

Coating Type	Water Vapor Permeance	
	ng/(Pa·s·m^2)	Perms
Modified asphalt 60 mils	0.2–0.3	0.003–0.004
Butyl 30 mils	0.8–1.1	0.015–0.02
Chlorosulfonate PE	11–17	0.2–0.3
Polychloroprene	11–17	0.2–0.3
Polyurethane	17–170	0.3–3.0
Acrylic	110–170	2.0–3.0
Silicone	160–170	2.8–3.0

- environment in which coatings are used (temperature extremes, wind potential, hail resistance, proximity to chemical emissions, risk for pedestrian abuse, etc.)
- aesthetic qualities (color, surface profile, reflectivity, and visibility)
- ability to withstand foot traffic (foot traffic resistance, rooftop HVAC servicing, and maintenance traffic)
- life expectancy (in relation to the design life of the roof system)
- ease of maintenance (drying time, ease of repairs or modifications with in-house maintenance personnel, ability to rejuvenate the coating system by re-application)
- field track record of use or laboratory testing pertaining to the application in question

1.4 Resistance to Hail Damage

Specific considerations must be given to coatings on roofs in areas with a high risk of hail damage. Hailstorms are selective in nature, hitting some areas and not others. Hailstorms are characterized by a variety of conditions including different hailstone sizes, speeds, angles, shapes, densities, and surface temperatures. The cool downdraft of air accompanies falling of hailstones to the earth's surface under the thunderstorm conditions. The cool air, the first wave of smaller hailstones, cold water on the surface and accompanying winds (speeds ranging from 40–60 mph) all have a cooling effect which lowers the roof temperature. Since the physical properties of polymeric materials are temperature-dependent, the impact of hail on cold surfaces increases the amount of damage to roofing systems.

Kashiwagi and Pandey (1997) examined performance of SPF roof systems with elastomeric coatings both in the laboratory, using the Factory Mutal Severe Hail (FM-SH) test, and in the field. The FM-SH test involves ten ball drops. If any of the drops results in breakage of the elastomeric coating, the system fails the test. Thirteen roofs with documented "severe" and "over-sized" hail damages were selected for evaluating the field performance of different types of elastomeric-coated SPF roof systems. There is a good correlation between the FM-SH test and field performance records.

The discussed research concluded that:

1. The FM-SH test had a good correlation with field tests on roofs exposed to severe hail damage.
2. Out of three tested elastomeric coated SPF roof systems, the polyurethane-coated SPF roof showed the highest hail resistance. It may be classified as a performing system for severe hail exposure. More specifically:

- One urethane (U2), 45 mils-thick coating system applied on 48 kg/m³—(3.0 pcf) density and 345 kPa—(50 psi) compressive strength SPF, performed without damage at both freezing and subfreezing temperatures.
- Another urethane (U1), 45 mils-thick coating system applied on 48 kg/m³—(3.0 pcf) density and 345 kPa—(50 psi) compressive strength SPF, performed well at freezing temperatures and not at subfreezing temperatures.

3. One acrylic coating, with 45 mil DFT (20 mil minimum), applied on 48 kg/m³—(3.0 pcf) density and 345 kPa—(50 psi) compressive strength SPF, passed testing at room temperatures, however, it failed under freezing temperatures.

4. This study showed that many acrylic coatings were applied below the required DFT. One must ensure that these coatings are fabricated in accordance with the specified minimum DFT.

5. The cooling of samples generally decreases the hail resistance of elastomeric coatings.

6. Granulation did not improve hail resistance of the tested coating system.

7. The accelerated test of UV exposure (1,000 hours) had no significant effect on hail resistance.

2. GENERIC TYPES OF ELASTOMERIC PROTECTIVE COATINGS

The elastomeric coating system may be one or more of the following types:

- acrylic
- butyl
- chlorinated synthetic rubber[27]
- silicone
- polyurethane
- modified asphalt

2.1 Acrylic Elastomers

Acrylics are single components. Most are water based which makes them easy to apply and easy to clean. They can be tinted to any color. Being water based, they must be protected from freezing, and they should not be applied

---

[27]These coatings are frequently referred to in the trade as "Hypalon" or "Neoprene" coatings. The terms "Hypalon" and "Neoprene" are registered trademarks of E.I. du Pont de Nemours & Co., Wilmington, DE.

when the temperature is lower than 10°C (50°F). They have high water absorption (one magnitude higher than other coatings) and low abrasion resistance and poor resistance to solvents and fair resistance to other chemicals. The water vapor permeance of these coatings is higher than the SPF and permits vapor "breathing" though the coating layer. Acrylics are generally applied in a DFT up to 15 mils. They must be applied in two coats to achieve the minimum DFT of 30 mils. Their curing time, depending on weather conditions, ranges from four to eight hours.

2.2 Butyls

These are two component coatings based on polyisobutylene polymers. Colors available are black, gray, or tan. Butyls are vapor retarders, i.e., they have permeance much lower than that of SPF. Butyls may have a low UV resistance. Therefore, they are often used for a basecoat. They are applied in a DFT between 10 and 15 mils, which requires a two–coat application to achieve the minimum DFT of 20 mils as the basecoat. The topcoat applied on top of a butyl may be either silicone, chlorinated polyethylene, or urethane.

Butyls have average mechanical properties, poor resistance to solvents and excellent resistance to other chemicals. Their curing time, depending on weather conditions, ranges from one to six hours.

2.3 Chlorinated Synthetic Rubber

2.3.1 CHLOROSULFONATED POLYETHYLENE (HYPALON ®)

Chlorinated polyethylene is a single-component coating. These coatings, known for their good fire performance and UV resistance, because of high solvent content, must be applied in thin coats. Their vapor permeability is slightly lower than that of SPF, though not as low as that of butyls. They have high tensile strength, and medium impact resistance, poor resistance to solvents, and good resistance to other chemicals. Generally applied in a DFT up to 8 mils, they must be applied in several coats. Their curing time, depending on weather conditions, ranges from two to six hours.

2.3.2 POLYCHLOROISOPRENE (NEOPRENE ®)

This is a single-component coating with performance similar to the other chlorinated synthetic rubbers. It may also need to be covered by a more UV-resistant coating.

2.4 Modified Asphalts

These coatings are made by adding rubber-like polymers to asphalts. Most of them cure by means of chemical reactions and not by solvent evaporation. Most are considered vapor retarders. Note that the modified asphalt will not adhere to other asphalts, but will adhere well to other coating systems.

2.5 Urethane Elastomers

Urethanes can either be "catalytically cured" (two-component) or "moisture cured" (one-component). They range from high solvent-content to solvent free, from vapor-retarders to vapor-breathers. They are available in either black or in many colors. An additional description is needed to define physical properties of a specific product.

Aliphatic urethane coating may be designed so that it retains gloss and resists "chalking." Aliphatic urethanes have high tensile and tear strength, medium to high impact resistance, and good abrasion resistance.

Aromatic urethanes have medium to high tensile and tear strength, high impact resistance, and good abrasion resistance. Aromatic urethanes are less UV stable than aliphatic urethanes and often are used as basecoats and are covered with either Hypalon, aliphatic urethane, silicone, or a UV resistant, aromatic urethane.

Polyurethane coatings may also be modified to include additives. These additives may either modify physical properties of the coating or be used to reduce price. Urethane coatings are applied in a DFT up to 20 mils.

2.6 Silicones

A silicone can be a single- or two-component coating, which forms a "breathable" membrane. Usually they are white or gray. Often they are used with colored granules. They have the highest heat and UV resistance amongst all types of elastomeric coatings. They are applied in a DFT up to 15 mils, i.e., applied in two coats. Their curing time, depending on weather conditions, ranges from one to two hours.

3. LONG-TERM PERFORMANCE OF COATINGS IN THE FIELD STUDY OF ROOFING SYSTEMS[28]

A number of field studies were performed by the Naval Civil Engineering Laboratory. They included

---

[28]This chapter is based on Alumbaugh (1993).

- the evaluation of SPF roofing systems at the Naval Reserve Center (NRC) in Clifton, New Jersey (Keeton and Alumbaugh, 1977, Alumbaugh et al., 1983)
- investigations of SPF aging (Zarate and Alumbaugh, 1982, 1986)
- fire performance of SPF applied to metal decks (Alumbaugh et al., 1983)
- guidelines for SPF roof maintenance (Alumbaugh et al., 1984)
- performance of SPF roofs in tropical environments (Alumbaugh et al., 1986)
- performance criteria for SPF roofs (Alumbaugh and Keeton, 1977, Alumbaugh and Humm, 1990)
- guidelines for SPF roof applications (Coultrapp et al., 1987)

Of these studies, only the evaluation of SPF roofs at the NRC Clifton will be discussed here.

The standing-rib metal roof of buildings at the NRC Clifton developed extensive leaks. A decision on SPF application was accompanied by a research program to determine the long-term properties of both SPF and elastomeric coatings (Alumbaugh et al., 1983). There were two parallel metal buildings connected by a concrete block structure with a wooden roof deck. The front metal building had dimensions of 48 m × 12 m (160 ft × 40 ft) and the second building was 60 m × 12 m (200 ft × 40 ft). After all loose materials were removed and the surface was cleaned, the primer and SPF with average density of 37 kg/m³ (2.3 lb/ft³) were applied. The roof area was divided into five sections, and initially five different elastomeric coatings were applied to these sections. In each case, the basecoat of the elastomeric coating system was applied the same day as the foam.

System 1. Catalyzed Silicone Rubber. This system consisted of a basecoat (medium gray) and a topcoat (cement-gray), which together gave a total DFT of 0.76 mm (30 mils). Half of the area with this coating had ceramic roofing granules cast into the wet topcoat.

System 2. Moisture-Curing Silicone Rubber. Two coats of this single-component coating had a DFT of 0.43 mm (17 mils). The basecoat was light gray and the topcoat was white.

System 3. Catalyzed Butyl-Hypalon. This system consisted of a two-component black butyl basecoat and a two-component white Hypalon topcoat to give a total DFT of 0.28 mm (11 mils).

System 4. Hypalon Mastic. One coat single component white Hypalon mastic was applied with a DFT of 0.81 mm (32 mils). (This material was difficult to apply.)

System 5. Catalyzed Butyl-Hypalon. This system consisted of one coat of a two-component, tan butyl basecoat and a one-component Hypalon topcoat to give a total DFT of 0.66 mm (26 mils).

Table 28. Performance rating of the SPF roofing systems
at Naval Reserve Center, Clifton, NJ.

System Number	6 m	9 m	18 m	2 y	3 y	4 y	5 y	6 y	7 y
1ᵃ	E	E	E	E	E	E, VG	E, VG	E, VG	E, VG
1ᵇ	E	VG, E	VG	VG	VG	VG	VG	VG	VG
2	E	VG, E	VG	VG	VG	G, VG	G, VG	G	F, G
3	G	F	P, F	P	—	—	—	—	—
4	VG, E	G, VG	G	P, F	—	—	—	—	—
5	VG, E	G, VG	VG	F, G	—	—	—	—	—
6	—	—	—	—	VG, E	VG, E	VG, E	G, VG	F, G
7	—	—	—	—	VG, E	VG	G	F, G	P, F
8	—	—	—	—	E	VG, E	VG, E	VG	VG

ᵃDenotes coating with granules.
ᵇDenotes coating without granules.

Early failure of coating systems 3, 4, and 5, which was partly due to poor quality of systems 3 and 4 and partly due to hail damage, necessitated recoating after two years. After cleaning the roof surface to remove any dirt, loose coating, or chalking, three other coating systems were applied.

System 6. Moisture-Curing Silicone Rubber with Granules. Except for white ceramic roofing granules applied in the wet topcoat at the rate of 2.4 kg/m² (50 lb/100 ft²), system 6 is identical to system 2. The basecoat and topcoat together with the old system 3 gave a total thickness of 0.60 mm (26 mils).

System 7. Catalyzed Hydrocarbon-Modified Urethanes. This system consisted of a black basecoat and a topcoat. The DFT of the old system 4 was 1.27 mm (50 mils). The total thickness of systems 4 and 7 was 2.03 mm (80 mils).

System 8. Catalyzed Urethane with Granules. This system consisted of a urethane primer followed by a basecoat of aluminum aromatic urethane (catalyzed) and a topcoat of white aliphatic urethane (catalyzed). White ceramic roofing granules were applied in the wet topcoat at the rate of 2.4 kg/m² (50 lb/100 ft²). Total dry thickness of the film together with the old system 5 was estimated at 1.2 mm (47 mils).

The performance of the SPF roofing systems was determined by semi-annual inspections for the first two years and annual inspections for the next five years. Table 28 lists performance ratings after specified periods of exposure:

- E means excellent—basically no coating and foam deterioration
- VG means very good—only minor coating or foam deterioration
- G means good—coating or foam deterioration nearing the point of significance
- F means fair—moderate deterioration

- P means poor—numerous areas showing moderate to severe coating or foam deterioration

Deterioration of the SPF system can only be noticed when the coating fails to perform. As long as the coating performs well, so does the foam. When the coating deteriorates or flakes off, the foam is exposed to weather and may deteriorate, though it may take quite a long time to have a thick layer of SPF deteriorate to such a degree that it cannot be repaired. No deterioration happened during the reported 7-year period; even the foam with the surface damaged by bird pecking or hailstorms did not show any moisture leaks.[29]

3.1 Catalyzed Silicone

Somewhat better performance was noted for the section covered with granules for the first five years. At the end of seven years of weathering, however, the areas were performing equally and both were rated very good.

Coating deterioration consisted primarily of cracking and flaking of the coating along the foam covering the standing ribs. One should note that the foam was sprayed only in one direction, rather than each pass of the foam being sprayed perpendicular to the previous pass (crosshatching). There appears to be equal amounts of bird pecking in both granulated and non-granulated areas, indicating that granules were not effective as a protective measure against birds.

3.2 Moisture-Cured Silicone

System 2 did not perform as well as system 1; it was probably more prone to the damage from foot traffic due to its smaller thickness. In areas where foot traffic was heavy, the foam was compressed, damaging the cell structure, and in some cases, loosening the coating from the foam. Furthermore, in one small area, the foam was wet. The overall rating after seven years was fair to good.

3.3 Butyl Basecoat and Two-Component Hypalon Topcoat

The topcoat of system 3 was very thin, which, together with poorer coating quality resulted in heavy hail damage after only one year and severe checking, cracking, and erosion after only two years of weathering. This system was rated as poor and was re-coated with system 6.

---

[29] A well-known case is that of New Orleans's dome roof damaged by a strong hailstorm. The dome was repaired only after a number of years of unprotected exposure.

3.4 Hypalon Mastic

One-coat, single-component Hypalon mastic of system 4 was severely damaged by hail and badly eroded on the south side of the roof. After two years, this system was rated as poor and was re-coated with system 7.

3.5 Butyl Basecoat and One-Component Hypalon Topcoat

This system (5), was performing well until it was damaged by hailstorms. After two years, this system was rated fair to good and was re-coated with system 8.

3.6 Butyl/Hypalon Basecoat and Moisture-Cured Silicone Topcoat

Applied over the existing butyl-Hypalon coating that had been badly cracked by hailstones, the silicone coating was not sufficiently thick to bridge these cracks. Nevertheless, the hydrophobic nature of the coating prevented water entry into these cracks.

Other forms of deterioration in this system consisted of spalling of the coating, minor bird pecking, and one spot of foam delamination (Alumbaugh and Humm, 1990). After five years, this system was rated very good to excellent. With time, flaking of the coating became progressively worse, and the system was rated fair to good.

3.7 Butyl/Hypalon Basecoat and Hydrocarbon-Modified Urethane Topcoat

The black basecoat and aluminum-filled topcoat, applied over the existing Hypalon coating appeared to bridge hailstone cracks, but did not bridge areas where the old coating was spalled. In addition, this coating exhibited heavy pinholing and moderate blistering. The system was rated good after five years and poor to fair after seven years of exposure.

3.8 Butyl/Hypalon Basecoat and Urethane Topcoat

System 8, applied over the existing butyl basecoat and Hypalon topcoat, was performing well and was rated very good after seven years of exposure, even though this system did not bridge the hailstone cracks in system 5.

4. EFFECT OF THE COATING ON LONG-TERM THERMAL RESISTANCE OF THE FOAM

After five years, samples were taken from each section of the tested roofs and their thermal resistivity was measured under standard laboratory conditions (ASTM method C518 at 24°C). Results are given in Table 29. As discussed in the Introduction, increasing the foam thickness and providing some additional means to slow air ingress into the foam would improve the time-averaged thermal performance of the foam. Table 29 illustrates the significance of coatings on the LTTR of the SPF. Since the metal roof substrate and the sprayed foam are identical in all cases, one may easily assess the significance of the coating system on thermal performance.

Systems 1 and 2 are thin and permeable for water vapor and air components. As judged by the measured thermal resistivity values, their aging rate is practically not affected by the presence of the coating.

Conversely, coatings 7 and 8 comprise different layers built to thicknesses of 1.2 to 2 mm. Table 29 shows that their ability to retard the aging of foam is evident. Even though the multiple coating was built only two years after SPF was applied, there appears to be a significant retardation of the aging process in roofs 7 and 8.[30] The thermal performance of SPF covered by system 8 is practically the same as gas-barrier-faced polyisocyanurate board (Sherman, 1981).

Table 29. Thermal resistivity determined after five-year aging on an experimental roof (Alumbaugh and Humm, 1990).

Coating Code	Coating Thickness (mm)	Vapor Permeable	SPF Thickness (mm)	r-Value (m K)/W	r-Value (h °Ft²)/(Btu in.)
1[a]	0.50	Yes	51.0	38.8	5.60
1	0.43	Yes	57.2	39.0	5.62
2	0.28	Yes	60.3	39.9	5.75
6	0.60	No	63.5	43.9	6.33
7	1.27	No	69.8	48.5	6.99
8[a]	2.03	No	63.5	49.2	7.09

[a]Denotes coating with granules.

[30]Use of a polymeric gas barrier to reduce the aging process of polyurethane is not new. Applying selected priming to slightly wet plywood surfaces in manufacturing of polyurethane skin panels proved to be effective (private communication with Herb Nadeau, Niagra on the Lake, 1983).

5. TESTING PHYSICAL PROPERTIES
OF ELASTOMERIC COATINGS

The following tests are used to assess the physical properties of elastomeric coatings used for roofing applications.

Tensile strength (ASTM D-412): The maximum tensile stress applied during stretching of a specimen to its rupture point. The type of die used, temperature, and speed at which the sample is tested should be reported.

Ultimate elongation (ASTM D-412): The maximum extension or stretching of the membrane at the time of rupture. The type of die used, temperature, and speed at which the sample is tested should be reported.

E-modulus (ASTM D-412): The ratio between the stress required to stretch a membrane and its elongation within the elastic region of stress/strain relation.

Tear strength (ASTM D-624): The measurement of the force required to propagate a tear in the membrane. Two methods are used to test the membrane, with a circular cut or with a 90° angle cut. The latter is a more severe test and will produce lower results.

Hardness (ASTM D-2240): A measurement of a membrane's inherent resistance to indentation. In many instances, a coating with a greater degree of hardness (in the Shore A range of 40–90) will have better abrasion resistance as well as resistance to cutting and tearing. Also, some harder coatings have better dirt release properties than softer coatings.

Abrasion resistance (ASTM D-4060): The measurement of the amount (by weight) of coating lost when subjected to an abrasive wheel of a Taber Abraser. The test is normally performed with a 1,000 gram weight and the weight lost is reported after 1,000 revolutions. The abrasive wheels used for elastomer membranes are either the CS-17 or CS-10 (the CS-17 wheel is more abrasive than the CS-10 and will produce higher weight loss figures). The type of wheel used should be reported. The abrasion resistance is related to the wear resistance and may be related to wear under traffic.

Impact resistance (ASTM D-2794): This test involves a procedure for rapidly deforming, by impact, a coating film and its substrate. The apparatus for this test is a cylindrical weight, which is raised and dropped within a tube onto the coating film from various heights. Failure is indicated by cracks in the film.

Water vapor transmission (ASTM E-96): The measurement of the rate of vapor transfer through a material or membrane.

Water absorption (ASTM D-471): The amount of water absorbed by a membrane when totally immersed in water at a given temperature.

Low temperature flexibility: There are two methods using an unsupported film of the membrane. ASTM D-2137 determines the lowest temperature at which flexible elastomeric materials subjected to impact will not exhibit fracture or cracking. ASTM D-2136 determines the ability of rubber-like materials, subjected to bending, to withstand low temperatures.

High temperature resistance (ASTM D-573): The maximum temperature a membrane can be exposed to without permanent deterioration. This information is necessary in applications where a coating may be exposed to abnormally high temperatures.

Heat aging (ASTM D-573): The resistance of a membrane to degradation when subjected to various specified elevated temperatures. A typical test temperature for an elastomeric coating designed for roofing is 70°C (160°F) for a minimum of thirty days.

Chemical resistance (ASTM D-471): The ability of a membrane to retain physical properties when subjected to spill, splash, or immersion conditions in various chemical solutions.

Accelerated weathering: Accelerated weathering can be used when membranes are subjected to an intense concentrated ultraviolet light, high humidity or condensation, and elevated temperatures. Two methods are commonly used to screen coatings for comparison purposes.

- ASTM D-822: an ATLAS Weatherometer, which normally cycles between a special light source and a water spray. The light source, which is correlated with the spectrum of the sunlight, can be either carbon arc or xenon. The carbon arc normally weathers the samples slightly faster than xenon, while xenon more closely simulates sunlight.
- ASTM G-53: a QUV Accelerated Weathering Tester. Special UV fluorescent bulbs are used as the sources of light. The tester is constructed to induce moisture condensation with temperatures up to 80°C (180°F) on the panel rather than using a water spray. Signs of film deterioration and retention of physical properties are recorded at various intervals.

Ozone resistance (ASTM D-1149): The resistance of an atmosphere containing a high concentration of ozone.

Mold and mildew resistance (ASTM D-3273): The resistance of a membrane to mold and mildew growth and the resulting deterioration of film integrity. In addition, the growth also causes an unsightly appearance of the finish.

Note: Tests related to performance under fire conditions are discussed later.

6. QUALITY ASSURANCE

6.1 Manufacturer's Responsibility

The manufacturer's responsibility is to provide a product that conforms to its claims relative to basic product description and uses, physical properties, and field performance. The manufacturer should provide the literature, QA in the manufacturing plant that includes shipping and handling, training and approval of applicators, and even a periodic inspection of their jobs.

Literature: Should include product description, basic uses, wet physical properties, cured physical properties, performance characteristics, fire rating and approval, building code and insurance acceptance, application instructions and techniques, limitations, and precautions.

Plant quality assurance: Manufacturers should insure batch-to-batch uniformity and determine that product quality is indeed within the established parameters. Manufacturers should also retain liquid samples, taken from each batch produced for a specified time.

Shipping and handling: The product should be packaged in clean, properly sealed and labeled containers according to ICC regulations and other pertinent laws. Coatings that are beyond their advertised shelf life should not be sold by manufacturers or distributors.

Applicator training and approval: Most manufacturers will help the contractor train personnel to handle and apply their products. This training can be undertaken in seminar-type programs or as an in-field exercise, depending on the complexity of the product and/or the equipment necessary for its application. Manufacturers may also require formal training prior to the sale of products or the issue of a license or approval of a specific application.

Job inspection: For warranted applications, many manufacturers require various inspections. The job should be visually inspected to determine that the following areas are in compliance with the manufacturers printed instructions: surface texture, uniformity of coating coverage, minimum coating thickness, existence of pinholes, evidence of non-coated polyurethane foam and overall appearance. Where deficiencies exist, these should be brought to the attention of the contractor for correction.

Coating DFT is usually measured from a slit sample using an optical comparator. After the coating is cured, a slice (slit sample) of about 25 mm (1 inch) long, 13 mm (1/2 inch) wide and 13 mm deep is taken from the surface of the foam. It is used to check the surface adhesion and the minimum DFT. The average of three readings taken on the peaks of high points of the surface texture is called the minimum DFT.

6.2 Contractor's Responsibility

The contractor should have knowledge of the product and its use and assume responsibility for its handling and proper application.

Knowledge of product: Contractors and their crews should be aware of all the parameters regarding the proper application of a particular product, including uses, packaging, mixing, storage, and all application requirements. To this end, the field personnel should be provided with the proper training and knowledge by both the contractor and supplier to successfully apply the particular system.

Equipment: Applicators must have a complete understanding of their equipment and its use with the particular material being applied. Of particular importance are mix ratios, solvents, pressures, output, filters, spray tip size, and operating temperatures, proper maintenance, repair, and clean-up of equipment.

Job inspection: Quality assurance is the responsibility of everyone involved; from the selection and testing of raw materials to the inspections of project slit samples. It is incumbent on all those involved to have the knowledge, equipment, and personnel to provide the most successful application possible.

Specifically, the following should be monitored: output measured in liters (gallons), wet and DFT, and areas covered.

7. FIRE PROTECTION AND FIRE TESTING

7.1 Introduction

When exposed to the flame of cutting torches, unprotected SPF will ignite, resulting in a flash fire. The burning will be brief, forming a layer of less flammable surface char. This initial burning produces combustible gases and black smoke. In confined interiors, combustible gases can accumulate and ultimately ignite resulting in flashover, a dangerous fire situation. Under these conditions, additional foam or other combustibles can become involved in the fire giving off additional combustible gases and feeding the fire. Polyurethane's contribution to fire becomes significant if the heat and gases are not dissipated and the temperature of the foam rises above approximately 380°C (700°F). Then, the surface char will no longer be able to protect the foam and the foam will fuel the fire as it degrades under these extreme temperatures.

To extend the time at which the foam would reach its auto-ignition temperature should a fire originate from other sources, building codes require that SPF and other cellular plastics be protected by thermal barriers. Building codes treat differently the exterior applications of SPF, such as roof systems, where combustible gases can easily dissipate, therefore, SPF is less likely to become involved in flash fires.

This section provides an overview of different tests used to evaluate fire performance of thermal barriers. Nevertheless, to assist in evaluation of the efficiency of fire retardants used during formulation of SPF, some tests, e.g.,

ASTM E-84 (Test for Surface Burning Characteristics for Building Materials), are also performed on the SPF layer alone. One must bear in mind that these ratings are used solely to describe the response of SPF to controlled fire conditions and are not intended to reflect hazards under actual fire conditions.

7.2 Definition of Thermal Barrier

A thermal barrier is a designed to slow the temperature rise under its surface and to delay involvement of any underlying material in a fire. A thermal barrier limits the temperature rise under its surface to not more than 121°C (250°F) after 15 minutes of the standard fire exposure (i.e., exposure complying with the standard time temperature curve of ASTM E-119—Test Methods for Fire Tests of Building Construction Materials). This is equal to the performance of 12.7 mm (1/2 inch) gypsum board. Barriers meeting this criterion are termed a "15 minute thermal barrier" or are classified as having an "index of 15."

It is generally required in North America that a thermal barrier approved by the local building code be placed on all cellular plastics (foams) used in the habitable space. Thus, SPF cannot be applied to the interior of a building without an approved 15 minute thermal barrier as defined by the applicable building code. For SPF applied on the exterior of the structure (i.e., roof insulation), a thermal barrier is not normally required.

Many types of thermal barriers are available on the market today, including but not limited to:

- gypsum board
- spray-applied cementitious materials
- spray-applied fibrous materials
- various proprietary materials

The thermal barrier should have a currently valid building code certification that lists a report number and date. In some cases, a local building code official will allow the use of a thermal barrier which has been tested to the satisfaction of the official, but is not yet certified by a code agency.

Generally accepted tests for thermal barriers include:

- UL 1715 Fire Test of Interior Finish Material
- UL 1040 Insulated Wall Construction
- FM 4880 Building Corner Fire Test
- U.B.C. Standard 26-2 Test Method for the Evaluation of Thermal Barriers

7.3 Testing Fire Properties

Coatings based on organic compounds are subject to combustion and exhibit characteristics similar to all other combustibles under fire conditions. Additives are incorporated into coatings to inhibit ignition and/or reduce fuel contribution.

The risks involved with combustible roof coverings include flame spread across the surface of the roof or interior flame spread resulting from combustible vapors and liquids entering the building through seams, joints, and openings in the roof's substrate.

Although Model Building Codes do not have the force of law, they are good references since most state and local codes (which do have the force of law) are based on their provisions.

The following organizations in the United States evaluate coating systems through data submitted by coatings manufacturers:

- the Building Official and Code Administrators (BOCA)
- the International Conference of Building Officials (ICBO)
- the Southern Building Code Congress International (SBCCI)

Canadian Construction Material Center of the National Research Council, provides evaluation reports to indicate compliance with the National Building Code of Canada (a national model code) and the Underwriters Laboratory of Canada works in tandem with the Underwriters Laboratory of the U.S.

These data are comprised of literature, UL, ULC, and FM approvals and independent testing results. Acceptance of the coating system allows the system to be used in areas that must comply with specific building and safety codes.

To be classified, both the coating and polyurethane foam must meet standards and inspections of Underwriters Laboratories. The system classification denotes a specific foam with a specific coating at a specified slope. Typically, for a re-cover over an existing roof with a non-combustible deck, SPF with minimum 38 mm- (1 1/2 inch) thick foam will assume the SPF's construction rating. For combustible decks, the lesser of the existing roof or the foam roof rating, will prevail.

7.3.1 UL-790 (ASTM E-108, STANDARD TEST METHODS FOR FIRE TESTS OF ROOF COVERINGS)[31]

This test is applicable to combustible and non-combustible decks exposed to severe (class A), moderate (class B), or light conditions of the fire

---

[31]Known also as UL-790.

test exposure (class C). Normally, a composite system including SPF and coating is tested under conditions of fire simulated outside of the roof. The characteristics examined are the flame spread and the resistance to fire penetration through the roof deck during the intermittent flame exposure test.

7.3.2 UL-723 (ASTM E-84,[32] STANDARD TEST METHOD FOR SURFACE BURNING CHARACTERISTICS OF BUILDING MATERIALS)

This test provides a comparative rating, in ceiling position, and is applicable to exposed surfaces.

A 10-minute exposure to a standard fire source, provided at one end of a 24 ft- (7.32 m) long tunnel permits the establishment of the speed of the fire propagation on the material surface. This test ascribes a flame spread classification (FSC) number or the flame spread index (FSI) to the tested material. Asbestos cement board is given FSI of zero and red oak placed on the floor of the tunnel represents FSI of 100. Ratings for fuel contribution and smoke development are also established during this test.

UL-1256—This tests roof deck constructions and membranes for their resistance to interior fires. Numbers 136, 181, and 206 in the UL Roofing Materials and Systems Directory are SPF systems.

UL-263 Fire Resistance (P Rating)—This measures the ability of the roof ceiling and structural members to withstand interior fires for specified periods, i.e., one, two, or three hours. New construction designs for SPF are provided in the UL Fire Resistance Directory, Design No. P 733.

The Factory Mutual Engineering and Research Corporation tests the flame spread potential of building materials and assemblies related to their end-use applications. The fire tests include ASTM E-84 Standard Test Method for Surface Burning Characteristics of Building Materials, ASTM E-119 Standard Test Methods of Fire Tests of Building Construction and Materials, ASTM E-108 Standard Methods of Fire Tests of Roof Coverings, the FM building corner test, and the FM heat release rate calorimeter.

7.3.3   ASTM E-119 (STANDARD TEST METHODS FOR FIRE TESTS OF BUILDING CONSTRUCTION AND MATERIALS)

This standard is applicable to structural assemblies of buildings and other assemblies that constitute permanent integral parts of the finished building. This test prescribes a standard fire exposure (time-temperature curve) and

---

[32]Known also as UL-723 tunnel test.

evaluates fire endurance, i.e., the duration for which the assemblies will contain the fire and retain their structural integrity. To assist in this evaluation, temperatures of unexposed surfaces of floors, roofs, walls, and other partitions are measured. Typical acceptance criteria include period of time (e.g., 2 hour rating) during which there was no passage of flame, gases hot enough to ignite cotton, and a structural integrity test. The latter involves a hose stream specified as the fire extinguishing measure applied to the surface. The rise of temperature on the unexposed surfaces is less than 139°C (250°F).

FM approved systems (Class 1) are listed in FM's Approval Guide. Additionally, FM publishes Loss Prevention Data sheets, such as 1-57 Rigid Plastic Building Materials, which discuss acceptable uses of various materials and components.

A closing note must deal with the relation between the rating obtained under controlled fire conditions and the fire performance of the building construction. Many factors, such as fire loads, their distribution and surface characteristics, delivery of fresh air (usually related to geometry of the fire compartment and ventilation capability), thermal characteristics of the enclosure, and moisture content of materials as well as the air are known to influence the time-temperature relation during the fire. To simulate all those factors in a test or even in a computer model of field performance is too difficult. Therefore, one must consider the results of the fire test as one of many considerations in assessing fire performance of building construction and assemblies.

CHAPTER 5

Quality Management in Construction Process

INDUSTRY-WIDE QUALITY MANAGEMENT is recommended for all SPF systems used in construction. It means that all elements in the process design, fabrication, and field performance monitoring should follow a coherent and documented approach. Such an approach is needed since analysis (Lstiburek and Bomberg, 1996) showed that the design of building envelopes for environmental control requires a number of iterations. Each material must be examined with regard to its compatibility and interactions with the adjacent materials and components. How could one ensure that the process in which such a large number of considerations by different professionals results in a predictable final product?

One needs a tracking system that could check the linkages developed during the process of design and construction, and the linkage between the requirements initially postulated by the user to the verification of the completed system performance. To document the final system performance, we postulate inclusion of the commissioning tests in the design/construction process. Therefore a QA system, based on the ISO 9000 series, should be incorporated in the Performance Linkage model (Lstiburek and Bomberg, 1996a). The ISO 9000 series is already a part of the daily practice in many other industries, and construction should not be an exception. After all, most of the construction problems would be easy to fix if they were observed in the design stage, but are difficult and expensive to correct during the occupancy stage.

1. BIDDING AND PROCUREMENT PROCESS WITH DATABASE OF PERFORMANCE RECORDS

Facility managers facing a retrofit or a new project must consider the system performance with a view to minimum maintenance, repair and operational costs. Traditionally, the facility managers have used the "low bid" competitive award system to select a construction with the lowest initial cost. Since all submitted bids fulfilled the technical specifications, it was assumed that performance of all systems will always be above the "specified" minimum performance level. Such an assumption could be justified if the design specifications were written in performance terms and addressed both the short- and the long-term aspects of field performance. This, however, is an infrequent case.

Facility owners currently use a forward path to design. This approach implies that the facility owner (or his/her representative) analyzes the project and develops several conceptual designs. One design is then selected for developing technical specifications. The contractors then bid on the project, and the qualified low bidder is awarded the contract. The drawback to the forward path process is that the specification favors only one conceptual solution. The facility owner may not have considered an alternative that could be better performing and have lower life-cycle cost.

Most specifications documents are based on properties used to characterize products (materials) and the correlation between them and the actual field performance of the system is tenuous. Kashiwagi et al. (1994) give the following two reasons for poor long-term performance of many construction systems:

- lack of methods to separate systems with excellent field performance from those with mediocre field performance
- tradition of selecting construction systems in a "low bid" environment

To ensure high performance of the facility system, Kashiwagi et al. (1994) recommend that the facility owner should be able to:

- define the performance objectives for the construction
- use a system of information management based on a backward path for selecting the optimum contractor/facility system combination

Such an approach called, the performance-based procurement system (PBPS), was developed with the help of "fuzzy logic" (Kashiwagi, 1995). The PBPS integrates performance information on materials, systems and records of the contractor. The PBPS aims to deliver "the best available" performing contractors and systems at the lowest life-cycle-cost.

The PBPS, as shown in Figure 29, includes the following steps:

- The facility owner identifies the performance requirements.
- Contractors/designers consider alternative design solutions, each with accompanying costs and experience (performance records) from previous installations.
- Performance data is collected for all different and possible design solutions.
- A relative distancing model (a part of PBPS) is used to compare the design alternatives and select the best contractor/facility system proposal.

It is evident that identification of the performance objectives is an important step in the PBPS. For instance, the following performance objec-

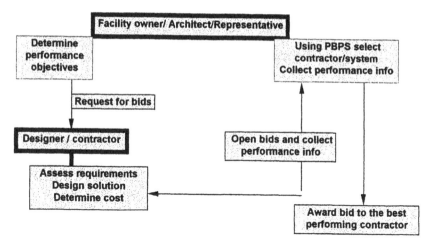

FIGURE 29. Performance-based procurement system.

tives[33] were used for the analysis of roofs (Kashiwagi and Moor, 1986):

1. Length of proven performance period
2. First cost and equivalent uniform annual cost (i.e., the simplest measure of life-cycle cost)
3. Percent of roofs not leaking
4. Percent of roofs not being maintained
5. Customer satisfaction
6. Frequency of traffic on a roof
7. The degree of disturbance to the operation of facility caused by the roof installation

In addition to a predetermined set of performance objectives, the PBPS approach requires information on contractors' experience (performance records) from previous installations. Then, using a weighted scheme that prioritizes the performance objectives and the relative distancing model, the PBPS, through a repetitive process, compares the design alternatives and selects the best contractor/facility system proposal.

2. PERFORMANCE IN MODELS OF THE DESIGN PROCESS

We use the term "model" as a synonym of knowledge. In any matter, as ex-

---

[33]This list may appear incomplete, however, the stress in Kashiwagi's research was on performance objectives that relate to the workmanship aspects of the construction process.

THE DESIGN PROCESS

1. ESTABLISH USER NEEDS
   FUNCTION OF THE BUILDING

2. ESTABLISH DESIGN REQUIREMENTS
   ENVIRONMENT
   SERVICES
   STRUCTURE AND ENCLOSURE

3. ESTABLISH DESIGN RESTRAINTS
   LEGAL
   TECHNICAL
   ECONOMICAL
4. MAKE BASIC DECISIONS ON FORM AND MATERIALS

5. MAKE DETAILED DECISIONS ON COMPONENTS AND CONSTRUCTION

FIGURE 30. Steps in the design process (from Baker, 1980).

pressed by Lord Kelvin, our understanding becomes much improved when we can quantify the relationships. Blach (1972), analyzing the question " how can performance be specified?" stated ". . . it is a necessary link both in development work and in situations calling for a choice to be able to *evaluate* the quality of solutions."

The word evaluate is underlined because the aggregate quality of a solution depends on many performance objectives. Even if some performance objectives can be easily specified, it ought to be the designer's task to decide which weight shall be given to different performance objectives. Furthermore, it must be noted that the great majority of test methods are not so accurate[34] that sharp distinctions between "pass/fail" categories can be justified.

The pass/fail criteria may be easier to establish when performance relates directly to the health and safety area, since traditionally this area is considered beyond the economic trade-off. Other aspects of performance are subject to a cost-benefit analysis. These trade-off aspects of the design/evaluation process require repeated iterations through the process models. We will illustrate it with a model presented by Baker (1980) for the design of roofs (Figure 30).

If the design requirements for the roof could be separated from those of

---

[34]In this context, "accuracy" means a judgement value for discrimination of the field performance (i.e., the correspondence of the result to the field performance).

other building subsystems, repeated analysis would be performed in steps 4 and 5 only. Otherwise, the whole process is repeated after any major change in design requirements caused by considerations in the other subsystems.

Baker's model represents design based on a "technology-driven" process where all technical considerations appear clear and coherent. Kashiwagi (1995) developed a PBPS for addressing the same issue from a logically opposite end. Instead of a forward path (user requirements, analysis to develop several possible solutions, screening the solutions and selecting one of them to develop technical specification), Kashiwagi selected a backward path. Kashiwagi argues that in the forward path approach, the facility owner may not have considered all alternatives and the specification clearly favors one solution. Since the screening is made in a conceptual stage of the design, there could have been an alternative and much better performing system (in terms of life-cycle cost) that was not considered in the design development stage.

These considerations have far-reaching consequences. Observe that the facility manager can formulate to only basic requirements without having a specific construction system in mind. This implies that during the design phase, two processes, the component selection and that the component evaluation are parallel. In effect, one may postulate that independently of how the design process is organized, it must allow fundamental changes through most of the design process duration. The design process (AIA, 1969) discusses five phases in the construction process:

1. Schematic design
2. Design development
3. Construction documents
4. Bidding or negotiation
5. Construction

It means that the fundamental changes in the design process should be allowed to occur until phase 5—until the actual construction is initiated.

To ensure that this is possible, we must change a design paradigm. A process model should be open to a continuous change, should highlight the interactions between design, environmental control, and performance objectives, and incorporate development of a quality assurance system. One such approach, called the circular performance linkage model (Lstiburek and Bomberg, 1996a), is discussed below.

3. QUALITY ASSURANCE DURING THE CONSTRUCTION PROCESS

This section reports the circular performance linkage (CPL) model (Lstiburek and Bomberg, 1996a). The term "circular" was introduced in the model's name to highlight that the design process is iterative (goes in circles). Thus, the four main stages of the CPL model that represent the main phases of the construction process are:

1. Conceptual design
2. Design development
3. Construction documents
4. Construction and commissioning

The last stage is called construction and commissioning because the quality assurance (QA) considerations warrant inclusion of commissioning into the design process. During each of these main stages, one must examine three main considerations:

• performance objectives
• design elements
• environmental control

These considerations of the CPL model are represented in Figure 31.

Finally, a QA system proposed for inclusion in the CPL model is based on the ISO 9000 series. The essence of the ISO 9000 system can be paraphrased as "write what you do and then do what you wrote." In other words, the main purpose of this QA system is to ensure consistency (repeatability and reproducibility) of the actions, without imposing any judgement value on these actions. (The judgement on the utility value of these actions is the responsibility of trained professionals). The presence of a formal QA system may alleviate some problems caused by a lack of communication between different groups of professionals and trades.

A scheme that prioritizes performance objectives should be documented as part of the design process. Furthermore, every time a major change was made either to the material or subsystem selection, performance information should have also been collected and documented. This would constitute a basis for QA, inspection, maintenance, and operational manuals.

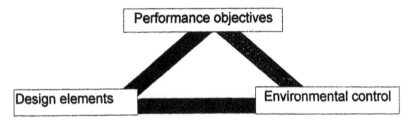

FIGURE 31. Main considerations of the CPL model apply for each of four stages.

Table 30a. CPL model: design considerations during the conceptual design stage.

Objectives (User Requirements)	Design Elements	Environmental Control
Site organization, building, elements of space separation, cost	Floor plans, facade, walls, roofs, floors, HVAC, structural system	Macroclimate (outdoor) and mezzoclimate of the building

Table 30b. CPL model: QA considerations during the conceptual design stage.

Objectives	Design Elements	Environmental Control
List user requirements and critical functions, weight of performance objectives	Establish design restraints: legal, technical, economical	Check orientation, winds, solar gains, mass effects, specific environmental requirements

Considerations for stage 1 (conceptual design) are listed in Tables 30a and 30b.

The design stage starts with an analysis of the user requirements needed to develop the project. They include elements such as: site selection and conditions, requirements for indoor space, facade, walls, roof, floors, HVAC system, and site subsystems (wind diversion, sun shading, etc.) which define environmental conditions for the building (mezzoclimate).

The AIA (1969) sets the following objectives for stage 1:

- illustrate possible solutions within the shortest possible time and at a minimum expense
- assist the client in understanding the program
- assist the client in determining the feasibility of the project

Similar to an Architect's project record, the QA manual should be initiated when the project is started. The original owner/user requirements, their weighted priorities and critical functions, and design restrains (legal, technical, economical) should be entered. At the end of stage 1, a formal check should be made of these environmental variables that are related to the site subsystems (wind diversion, building orientation, window and sun shading, mass effects, etc.) and define the building mezzoclimate.

Considerations for stage 2 (design development) are listed in Tables 31a and 31b.

When preliminary concepts and drawings are approved, the small team of architects working in stage 1 grows to involve structural designers, electrical, mechanical, and acoustical engineers, cost estimators, etc. The design development stage is the heart of the design process and the place where all subsystems are developed and all interactions should be examined.

The process of design starts with owner/user requirements. They are translated to specific, design and performance objectives which will assist the design team in attaining the specified construction objectives. As these performance objectives include both time-independent and time-dependent effects (durability and serviceability), it is important that the owner (owner's representative) also specifies the required service life for various subsystems. An understanding of the requirements and performance objectives as specified by the owner (user) is critical for the success of the design process. As the end of this process, commissioning and testing will be used to examine to what extent the design and construction satisfied the initial owner/user requirements.

A specific environmental control consideration is a design of the second line of defense. Theoretically, one can design and build a perfect structure. Yet, experience shows that in construction, eventually, something goes wrong. It may rain during construction time, or water may enter for other reasons, for instance, faulty flashing. For this or other reasons, some moisture finds its way into the envelope. So, as the second line of defense, walls are constructed to permit draining and drying of any excess moisture.

The second line of defense forces designers to consider the issues they can't predict, such as material changes or workmanship issues that escape their notice. Typically, the unnoticed leakage path often occurs in places such as: unfinished parging of masonry walls, behind radiator cabinets, or through holes used to pass wires to the suspended ceiling or corrugated roof decks. It also occurs when cladding materials are not drawn tight to

Table 31a. CPL model: design considerations during the design development stage.

Objectives	Design Elements	Environmental Control
Load, movements, strength, durability, fire protection, cost, etc. (see Table 1, part 1)	General assembly, all subsystems including security and communications	Subsystems forming indoor environment, second line of defense

Table 31b. QA model: considerations during the design development stage.

Objectives	Design Elements	Environmental Control
Record of the objectives and operational parameters of each subsystem	Criteria for design and selection of elements, components, and materials	List of critical parameterrs for indoor environment

Table 32a. CPL model: considerations during the construction documents stage.

Objectives	Design Elements	Environmental Control
Service life and life-cycle cost of all subsystems, review of quality manual	Construction details, material selection and buildability	Microclimate within the building envelope, subsystem interactions

Table 32b. QA model: considerations during the construction documents stage.

Objectives	Design Elements	Environmental Control
Record modifications to operational parameters of each subsystem	Check all construction details, location of HVAC ducts and grills	With computer models check microclimate within the building envelope

metal studs and through furred partition walls connecting with a suspended ceiling.

Finally, it is necessary to perform the cost-benefit analysis such as pay-back or life-cycle cost. AIA (1969) highlights this point: "Laymen generally have little comprehension of initial vs. ultimate cost, or their judgement may be clouded by the pressures of immediate budget demands. As the owner's professional advisor, it will be necessary for the architect to clarify these cost relationships during selection of materials, finishes and equipment."

Design development can either make the initial concepts more precise or change the design solution altogether. Such changes should be entered in the QA manual. Generally, objectives and operational parameters of each subsystem selected by the design team should be entered in the QA record. The QA records should also include criteria for the design and selection of elements, components, and materials as well as the list of critical parameters for the indoor environment. Specific points of environmental control in stage 2 are:

- check the continuity of environmental barriers
- incorporate the second line of defense

Considerations for stage 3 (construction documents) are listed in Tables 32a and 32b.

When design development documents are approved, the team of designers proceeds to review the construction cost estimates and develop the construction documents, i.e., working drawings and specifications. These documents will set forth in detail all requirements for the construction of the entire project. Development of working documents offers an opportunity of detailed control and verification of these construction details that usually cause problems and may contribute to failures. A plain area of wall or roof seldom originates the failure. The most frequent places of failure initiation are joints, junctions, and penetrations.

The clarity and quality of the design details are often the keys to good and long-term performance of the system. Perrault (1978) stated in a seminar on construction details for airtightness, "A good or bad detail will often make the difference between good and bad installation. Building designers should bear in mind that those who actually build the buildings usually have no design background. They should not be forced to guess the designer's intention, or to play the role of designer, but should only be expected to build carefully, as detailed. That is why a detail must be precise, easy to understand and predictable." Some specifiers prefer, therefore, to have one section of the specification addressing the joints and junctions of various subassemblies within the building envelope.

During this stage, the QA underlines the need for careful review (troubleshooting) of shop drawings of the assembled system as a whole and the design details. The significance of this review cannot be overstated. The specification terminology used on the design details is reviewed for clarity for those who rely on the drawings. (Note that often, jargon terms and abbreviations used by the designer are not understood by a tradesman.) The architect/ specifier may even choose to have a separate section addressing the joints and junctions of various sub-assemblies within the overall building envelope system.

Another critical issue is that of buildability of all design details. Like the second line of defense, buildability relates more to the judgment and knowledge than to mathematical analysis. Buildability reflects whether the specific design can be assembled by various trades without compromising the functional requirements during construction.

Contrary to a frequent misconception, buildability is more related to good design than to superior workmanship because, as experience indicates, only good design can combine all the environmental factors while presenting an easy construction pattern. For the most part, the designer must attend to the aspects of buildability such as material installation under different weather conditions, level of skill required for installation, and construction tolerances. Often, buildability problems arise when different professions are involved, for instance, neither the window manufacturer nor the wall designer are concerned with the window/wall interface. Yet, more wall durability problems were caused by the wall window/interfaces than by the window itself.

The same approach extends to the environmental control where the subsystem interactions are checked on a detailed level (e.g., tightness of ducts, connections, building envelope penetrations, etc). HVAC systems, location and size of ducts, elevator shafts, other mechanical equipment, illumination, and acoustics are these elements that require especially careful examinations.

Table 33a. CPL model: design considerations during the construction/commissioning stage.

Objectives	Design Elements	Environmental Control
Construction schedule, quality and cost	Inspection of subsystems	Exposure of materials to weather, rain, frost

Table 33b. QA model: considerations during the construction/commissioning stage.

Objectives	Design Elements	Environmental Control
Check if system performs to its original requirements	Inspection, verification and commissioning	Verification and testing for quality assurance

Specific points of design considerations in stage 3 are:

- check all construction details
- check all the penetrations
- check buildability of all design details

Specific points of environmental control in stage 3 are:

- use computer models to check the microclimate within the building envelope
- review HVAC/BE interactions

Considerations for stage 4 (construction/commissioning) are listed in Tables 33a and 33b.

The final stage of the construction process, called construction and commissioning (Tables 33a and 33b) is focussed on inspection, verification and commissioning of the subsystems and environmental conditions as specified by the QA manual.

As previously stated, the QA document is to provide the linkage between different professional concerns through all the design stages. Like a laboratory book with notes, comments, and references, the QA manual is supposed to outline the critical considerations affecting the subsystem selection and the performance objectives identified during the design process. These are considerations and requirements, which should be checked in the final commissioning stage.

Who is to perform these checks? Certainly, it is the principal architect in stage 1. The considerations during design development stage and their checks can be made by the design team. Stages 3 (construction documents) and 4 (construction/commissioning) requires, however, a special expertise in trouble shooting and testing. We shall denote this professional as a building science/diagnostics expert.

CHAPTER 6

Technical Support
for the SPF Contractor

INTRODUCTION TO THIS book postulated that the SPF industry should relate the QA system to the SPF field performance. To this end, the SPF contractor must be able to fabricate the most appropriate product with a view to achieving good, long-term, performance under specified service conditions. In doing so, the SPF contractor needs a strong technical support that involves the means for assessment of the product performance, review of its compatibility with adjacent materials, testing during the field fabrication and support during the follow-up (maintenance) programs.

The second and third chapters in this book reported factors affecting SPF performance and reviewed several laboratory tests for determination of physical properties of foam products. Normally, results obtained from these tests are reproducible and suitable for presenting the SPF products in the bidding documents. Some of these tests, however, are not suitable for predicting the end-use performance. Then, protective coverings and barriers were reviewed and the importance of the environmental control aspects for performing and lasting envelopes of buildings were highlighted.

So far, however, the tools aiding the SPF contractor in linking the quality assurance and SPF field performance were not provided. This is the role of the current section. To support the SPF contractor, the system house must deliver technical information that covers four areas:

1. Initial characterization of the SPF fabrication process
2. Basic laboratory tests of SPF physical properties
3. QA tests during the field fabrication of SPF
4. Equipment best suited for the application

The information defining the initial characterization of the fabrication process should contain:

1.1. Density (core, surface skins, knitlines, overall density)
1.2. Foam reactivity parameters (cream, tack-free time, maximum temperature)
1.3. Typical operating parameters and conditions of foam application
1.4. Comparative workmanship and SPF performance evaluation

The SPF contractor, together with the manufacturer, must define the acceptable range of variations (acceptable to both parties) for operational parameters critical to the SPF field fabrication process. Since operating parameters of the spray equipment may be adjusted depending on the application conditions, the physical properties of the foam may vary. The quality management concepts permit such variance, but require that the limiting values of operating parameters be defined prior to the actual foam installation. Developing this information will normally be combined with applicator training (see later text).

The basic laboratory tests of SPF physical properties should include:

2.1. Adhesion and cohesion (or tensile) strength
2.2. Key mechanical properties (compressive resistance)
2.3. Thermal properties (initial thermal resistance and LTTR)
2.4. Dimensional stability (maximum shrinkage and expansion)
2.5. Other properties as deemed necessary for the specific application

The information defining QA during field fabrication of SPF should contain the following tests:

3.1. Spraying a foam pattern and visual examination of the foam
3.2. Field determination of core density
3.3. Tack-free time and the maximum (exothermal) temperature
3.4. Adhesion and cohesion strength
3.5. Preparing the sample for external testing, when necessary

One may observe that the laboratory tests listed here are only a small fraction of the list normally contained in foam product standards. Several new tests and requirements increasing control during the foam fabrication are, however, postulated. Indeed, the system introduced here under the name of "Performance Based Quality Assurance" (PBQA) selects only the most important characteristics affecting the product's field performance and eliminates those with limited discriminating facility. The other properties will still be used by standard writing organizations, they, however, are not critical for achieving the required field performance.

Therefore, the PBQA requires one to:

• define the operational parameters critical to the SPF field fabrication process
• include them in the quality assurance process

The operational parameters may vary from application to application. The PBQA requires, however, that limits of operating parameters be defined prior to the field installation of SPF. In short, the PBQA system requires performing an initial characterization of the SPF fabrication process to establish these limits. Then, during each SPF fabrication, the contractor must perform some field tests to prove the compliance with these operational limits. In this manner, the contractor ensures that performance objectives of the construction project are fulfilled.

Test methods used for PBQA program are discussed below.

1. INITIAL CHARACTERIZATION OF THE SPF FABRICATION PROCESS

When purchasing the SPF from the system manufacturer, the PBQA requires the establishment of the limiting values of operating parameters, prior to use of the product in the field. The following section lists tests defining the SPF fabrication process.

1.1 Density (Core, Surface Skins, Knitlines, Overall Density)

Determination of core density is necessary in many situations involving the SPF contractor, such as system selection and its field fabrication, though different techniques are used in both cases.

The fundamental property of SPF is the core density, defined as the weight (in air) of a unit SPF volume. The core is obtained by removing 3 to 6 mm (1/8 in. to 1/4 in.) trim on both external surfaces. Descriptions of the sample preparation and the measurement of the core density are presented elsewhere in this Issue, namely in the section dealing with SPF testing.

Typically, three specimens, 305 mm × 305 mm- (12 in. × 12 in.) square with thickness of 25 mm (1 in.), are prepared for this test. These specimens are later cut into four smaller, 150-mm × 150 mm- (6 in. × 6 in.) square specimens, used for determination of compressive resistance. The apparatus and specimen conditioning procedure conform to the requirements of "Test method for apparent density of rigid cellular plastics" specified by ASTM standard D 1622.

Densities of surface skins and knitline areas are also important means to characterize the foaming process. Observe that the overall density is often used instead of the core density. It is even used for estimating the foam yield, which is incorrect. To establish foam yield one need the foam volume far in excess of that used for density measurements. The overall density is used[35] to classify the SPF types as well as in the initial development stages, when foam expander and hand mixing are employed, prior to use of the spray equipment.

---

[35]The minimum density for HD-type SPF is listed as 45 kg/m^3 (2.8 lb/ft^3). Nevertheless, the compressive strength required for HD-type, in the first chapter, is reached by SPF with density of 40 kg/m^3 (2.5 lb/ft^3). This review allows using 40 kg/m^3 (2.5 lb/ft^3) density foam in specific climatic areas and increasing to 40 kg/m^3 (2.5 lb/ft^3) in areas with extreme hot or cold climate.

1.2 Foam Reactivity Parameters (Cream and Tack-Free Time, Maximum Temperature)

The following parameters must be determined to define the foam reactivity: cream time,[36] tack-free time, maximum temperature and time of it occurrence.

The last test requires spraying about 40 mm- (1-1/2 in.) thick foam layer on a sheathing panel with thickness of 12 mm (1/2 in.) and area of 0.9 m × 0.9 m (3 ft × 3 ft). Then, immediately after spraying the first SPF pass, a temperature sensor (e.g., thermocouple) should be placed on the foam surface and fastened with a piece of tape. Start recording time. The second, 40 mm- (1-1/2 in.) thick foam layer, should be applied as soon as possible. Periodically, time and temperature should be recorded to extract the shortest period at which the maximum temperature was achieved.

1.3 Typical Operating Parameters and Conditions of Foam Application

Typically, the spray equipment is set within the following range of conditions:

- static primary mixing temperature, 43–60°C (110–140°F)
- static primary pressure, 550–830 kPa (800–1,200 psi)
- hose temperature, 35–54°C (95–130°F)

Typically, SPF is applied when the ambient temperature is in the range between 5°C (40°F) and 35°C (95°F). However, system houses offer seasonal grades of SPF formulated in relation to ambient or substrate or both temperatures anticipated during the time of application.[37] The manufacturer must be consulted in this respect.

As a rule, air humidity is less than 80%, however, it is practical to combine both temperature and humidity and require that ambient temperature be at least 3°C (5°F) above the dew point. The spraying should not take place, however, if there are any visible signs of moisture on the surface. Consideration should also be given to potential problems such as rainfall, mist, fog, ice, or snow.

Wind below 5.5 m/s (12 miles per hour) may be sprayed without windshields (providing that the risk of overspray is small). Enclosures are used for stronger winds, although winds above 12 m/s (25 miles per hour) may pre-

---

[36]See Appendix 2: Glossary.
[37]For application at temperatures beyond this range, consult the manufacturer.

vent outdoor application of the SPF.[38] Note that robotic spray application may be somewhat less sensitive to windy conditions.

Typically, two hours is considered sufficient time for curing[39] before applying the protective coating. It is necessary to consult the manufacturer, because some solvents used in coatings may damage green foam. It is preferable that the base coating being applied on the same day as the SPF application.

1.4 Comparative Workmanship and Foam Evaluation

The following example illustrates the application of PBQA model to a combined workmanship and SPF performance evaluation.[40] Three contractors (A, B, C) were given a choice of five roofing systems (I through V) and two spray guns (1, 2) to fabricate SPF for the comparative evaluation. The foam was sprayed on the substrate that was 2.4 m (8 ft) long and 4.2 m (14 ft) wide. The overlap between two passes was achieved on 0.9 m- (3 ft) to 1.2 m- (4 ft) distance. Practical observations made during the foam application were supported with the testing of local density and compressive strength.

The proportioning machine was Graco Foam Cat 400, fed by Gusmer 2:1 transfer pumps. Temperature and pressure were set in the middle of the operational range, namely at 49°C (120°F), and the pressure was set at approximately 690 kPa (1,000 psi). After applying a few passes, the temperature was dropped to 46°C (115°F). The following spray guns were used: Foam Cat with 421 module and Gusmer GX-7 a #1 module and 90 PCD (system I) or 100 PCD (for systems II through V).

Two contractors sprayed each of the tested systems and Tables 34, 35 and 36 show results for systems I, II, and III.

The following observations were recorded for product I: Contractor A—good reactivity, no creep, gun 1 gave smoother lap line than gun 2; and Contractor B—the foam stayed soft for too long time.

The results shown in Table 34 show that overall density manufactured with both guns is for all practical purposes identical. Nevertheless, the foam manufactured with gun 2 has much lower core density and much better green compressive strength (probably because it has better cellular structure). The knitline density and compressive strength obtained with gun 2 are higher than those obtained with gun 1, as well as the corresponding values for the core specimens made with either gun. The results obtained with gun 1 also support observations about smooth transition between different passes.

---

[38]For more detailed information consult the equipment/foam system manufacturers.
[39]Use manufacturer's recommendations.
[40]Data from Darrel Bennett, Foam Enterprises, Houston, private communication.

Table 34. Overall core and knitline densities and early compressive resistance results obtained during the comparative workmanship and SPF performance evaluation for the SPF product I.

Spray Equipment	Overall Density kg/m³ (lb/ft³)	Core Density		Compressive Resistance		Knitline Density		Compressive Resistance	
		kg/m³	lb/ft³	kPa	psi	kg/m³	lb/ft³	kPa	psi
Gun 1	42.9 (2.68)	40.4	2.52	185.3	26.9	42.1	2.63	208.1	30.2
		42.3	2.64	208.8	30.3	42.8	2.67	250.1	36.3
		41.7	2.6	186.7	27.1				
		42.6	2.66	183.3	26.6				
Average		41.7	2.61	191.0	27.7				
Gun 2	42.5 (2.65)	37.0	2.31	248.0	36	45.7	2.85	320.4	46.5
		39.1	2.44	234.9	34.1	45.7	2.85	331.4	48.1
		37.5	2.34	252.2	36.6				
		39.1	2.44	279.7	40.6				
Average		38.2	2.38	253.7	36.8				

There appears to be no significant difference between either density or compressive strength measured on the core and knitline samples.

Evaluation performed for product III is shown in Table 35. The following observations were recorded by the contractors:

C—good foam reactivity, no creep with gun 1 and very little creep with gun 2;

B—setup was good, the smooth ties on the lap line are similar to the SPF system. Both contractors rated this foam in a first place among the five candidates.

Table 35 shows that the overall density of product III manufactured with gun 1 was slightly higher than that manufactured with gun 2. The core density was slightly lower with gun 1, and the knitline density was slightly higher than that of the foam manufactured with gun 2.

The differences in compressive strength measured on the core and knitline samples appear to be smaller for gun 2 than for gun 1, while in both cases, the knitline specimens have much higher densities.

The results for product III (Table 35) show that the overall density manufactured with gun 2 was slightly higher (though close to that manufactured with gun 2), but the core density obtained with both gun 1 and 2 were identical. The knitline samples show higher density than those of the core samples, but their compressive strength is about the same as that measured on the core samples.

Evaluation performed for product V is shown in Table 36. The following observations were recorded by the contractors: C—slightly slower foam reactivity, the foam stayed soft longer time, and A—started good but surface became a little rough.

The ties on the lap line are similar to the SPF system IV where the overlap lines were seen on the 3- to 4-feet wide strip. Despite good results from compressive strength test, the contractors rated foams IV and V as the worst two among the five candidates.

Tables 34, 35, and 36 support the previous statement that the overall density is not suitable for comparing SPF products. However, the core and knitline densities combined with observations of creep and tests of compressive resistance are useful for product comparisons.

Experience obtained from the discussed comparative workmanship and foam evaluation permits us to propose the following process for standard use in the PBQA system:

1. Select three foam products for comparative evaluation. The contractors are given a form to be filled in during this evaluation and the check list which includes reactivity (tack free time), texture, sprayability, flowability, temperature sensitivity, application range, etc.
2. Perform the following QA tests during the SPF fabrication:
   - spraying a pattern

Table 35. Overall core and knitline densities and early compressive resistance results obtained during the comparative workmanship and SPF performance evaluation for the SPF product III.

Spray Equipment	Overall Density kg/m³ (lb/ft³)	Core Density		Compressive Resistance		Knitline Density		Compressive Resistance	
		kg/m³	lb/ft³	kPa	psi	kg/m³	lb/ft³	kPa	psi
Gun 1	41.5 (2.59)	38.0	2.37	290.8	42.2	46.0	2.87	308.0	44.7
		34.1	2.13	236.3	34.3	46.1	2.88	321.8	46.7
		38.1	2.38	250.1	36.3				
		34.1	2.13	241.2	35				
Average		36.1	2.25	254.6	37.0				
Gun 2	40.7 (2.54)	37.3	2.33	291.4	42.3	44.5	2.78	310.1	45
		37.6	2.35	304.5	44.2	44.9	2.8	276.3	40.1
		37.6	2.35	290.8	42.2				
		38.1	2.38	290.8	42.2				
Average		37.7	2.35	294.4	42.7				

Table 36. Overall core and knitline densities and early compressive resistance results obtained during the comparative workmanship and SPF performance evaluation for the SPF product V.

Spray Equipment	Overall Density kg/m³ (lb/ft³)	Core Density kg/m³	Core Density lb/ft³	Compressive Resistance kPa	Compressive Resistance psi	Knitline Density kg/m³	Knitline Density lb/ft³	Compressive Resistance kPa	Compressive Resistance psi
Gun 1	43.9 (2.74)	39.6	2.47	279.7	40.6	46.0	2.87	308.0	44.7
		41.8	2.61	329.3	47.8	47.7	2.98	353.5	51.3
		40.2	2.51	292.1	42.4				
		42.5	2.65	297.0	43.1	48.1	3.00	230.8	33.5
Average		41.0	2.56	299.5	43.5				
Gun 2	45.5 (2.84)	40.7	2.54	333.5	48.4	44.5	2.78	303.8	44.1
		41.2	2.57	349.3	50.7				
		41.0	2.56	342.4	49.7	44.7	2.79	303.2	44.0
		41.3	2.58	308.0	44.7				
Average		41.1	2.56	333.3	48.4				

- field determination of the core density
- maximum temperature and time of it occurrence

3. Choose one foam system and apply it under three different operating conditions, namely, at two levels of pressure (first at pressure lower, then at pressure higher than the standard conditions). Tests 1 and 2 are performed at the standard temperature. Then, maintaining the same pressure as Test 2 (higher than standard pressure) the temperature is lowered below the standard conditions and Test 3 is performed.

4. Cut three specimens for density and compressive strength determination from core and knitline of each of these three Tests (total 18 specimens). If the compressive strength testing is not performed at the foaming location, prepare samples for external testing.

5. As an option, one may perform two additional tests, namely: repeating Test 1 and 2 with another spray foam gun (or nozzle).

Performing the comparative workmanship and foam evaluation is very important to the contractor. It combines assessment of the foam texture, and ease of application with performance characteristics. By including some installation parameters into this training/evaluation, the contractor is provided with the knowledge of an acceptable range of foam product variations.

The PBQA system accepts such variations, but requires that the limits operating parameters be defined prior to the actual installation. Furthermore, the knowledge obtained during this training/evaluation routine permits adjustment of the operating parameters of the spray equipment, as needed, depending on substrate and environmental conditions.

2. BASIC LABORATORY TESTS OF SPF PHYSICAL PROPERTIES

This section lists the basic performance data, which should be provided to the SPF contractor by the system house.

2.1 Core Density

ASTM C 1029 specification explains that removing 3 to 6 mm from both the external skin and boundary skin formed at the substrate/foam interface is sufficient to obtain the core specimen. Core specimens may contain one or more internal skins at the spray pass boundaries. This specification refers to ASTM test method D 1622 for measuring the apparent density of the specimen. An alternative test method is defined in Section 3.2 of this chapter.

2.2 Adhesion/Cohesion Strength

This may be based on ASTM test method C 165 -92 or on the test defined in Section 3.4 of this chapter.

2.3 Compressive Resistance (Strength)

The compressive resistance is a load per unit of area at specified deformation, normally 10% of the original thickness. Therefore, the specimen thickness cannot be less than 25 mm (1 in.). This test is defined by procedure B in the ASTM C165 standard.

The mechanical properties must be determined during the initial characterization of the foam product for a case of a potential dispute involving both the system house and the SPF contractor.

2.4 Thermal Properties (Initial Thermal Resistance and LTTR)

The methodology is described in Chapter 3, and Appendix A1.

2.5 Dimensional Stability (Maximum Shrinkage and Expansion)

The methodology is described in Chapter 3, Section 3.3.

2.6 Other Properties That Are Deemed Necessary

As a minimum, to define a SPF product, one must use the five above-listed characteristics. They, together with field QA characteristics such as yield, reactivity, and core density, are necessary to outline SPF performance in a construction system.

Many more characteristics are listed in the technical literature, however, their use depends on the specific application. For instance, if the SPF is used in the engineering design, such as stress panels, one may need the following additional information:

- water vapor permeance
- mechanical performance under static and dynamic loads
- moisture accumulation under constant thermal gradient conditions
- maximum differential deformation of the foam

3. QA TESTS DURING THE FIELD FABRICATION OF SPF

A SPF installer should be qualified and licensed by a recognized third party accreditation program such as SPFD/SPI in the United States and CUFCA in Canada.

These QA programs normally require that, for each job site, the SPF installer perform the following operations:

1. Keep records: Work records of job site, date, material used, name of the installer, application conditions, temperature, results of thickness, and density measurements or adhesion/cohesion (when required). A sketch showing where material is used is also recommended.

2. Spray a test pattern. The spray pattern is to ensure that the proper equipment setting has been selected. The pattern shall be sprayed vertically with a gun positioned at a distance of about 1 m (3-1/4 ft). Consideration shall be given to reactivity, pattern of foam growth, and the appearance of the foam surface (see SPFD or CUFCA training manuals).

The PBQA system, introduced earlier in the book, further requires the following:

1. The field determination of core density should be performed on three specimens at the beginning of each day or at each time the material batch or job site are changed.

2. If any of the attributes observed during the spraying of the test pattern is questionable, the maximum temperature rise test should be performed. If either density or the maximum temperature are beyond the limiting values, then the sample should be prepared and sent for external testing (see the next chapter).

3. Adhesion and cohesion strength test is optional for new construction, however, the PBQA that this test be performed for all re-covering and re-roofing applications.

3.1 Spraying a Test Pattern

An experienced operator is able to check some foam attributes during fabrication of the spray pattern. These attributes are: color, texture, surface tack (speed of reaction), cell size, surface friability, and uneven growth of the foam. Some of these considerations, which are discussed in detail during the SPF contractor training are listed below:

Color and texture. While color may vary from one system to another, any change in the usual color may indicate a problem (e.g., poorly mixed chemicals because of poorly operating transfer pumps, screens clogged, etc. Light-colored and soft foam may indicate lack of A-component, dark and brittle may be caused by lack of B-component.) Standard trouble-shooting techniques should be used.

A sample of foam must also be taken and the core must be carefully examined for any discoloration that may be indicative of scorching. (Scorching may be caused by passes that are too thick high or by temperatures of the primary or the hose heaters that are too high).

Surface texture variation (e.g., popcorn/treebark textures, see Appendix A3) may also be indicative of a mixing or application deficiency.

Speed of the reaction (surface tack). Foam reactivity depends on many factors: chemical composition of A and B components, temperature and humidity of the ambient air, temperature and type of substrate, temperature of primary and hose heaters, dynamic and static pressures on the spray gun, etc. After extensive use of a system, most installers become familiar with the reaction speed under given set of installation conditions. Any deviation from the normal reaction speed should be viewed with alarm and the manufacturer should be contacted.

A particular case when the SPF surface that remains tacky may indicate poor mixing, improper system selection, too low temperatures on the primary and hose heaters or too-thin pass applications. Unless the equipment is still in the warming stage, the operating conditions should be adjusted and the test pattern repeated.

When test pattern produces results different from the expected, a cohesion/adhesion test should be performed.

Cell size. A fine cell size is expected. Coarse cell structure (e.g., the average cell diameter higher than 0.5 mm) may reduce both mechanical and thermal performance and dimensional stability.

Coarse cell structure may be caused by poor mixing, contamination of chemicals by oil or water, or surfactant failure in the SPF system. Check the machine and sources of contamination. If the cause cannot be found, prepare the sample to be sent to the chemical manufacturer.

Friability (brittleness). When running a fingernail over the surface one may sense a presence of fine powder. Since an initial friability may be associated with a curing process, the same operation should be repeated two days later. (Excessive friability may also indicate that the SPF system is not suitable for a low application temperature.).

Thermal cracking. Loss of adhesion, particularly during cold weather spraying, is sometimes characterized by a sharp cracking noise. Thermal cracking is usually caused by improper system selection and may be indicative of reduced adhesion and dimensional stability of the foam. The adhesion/cohesion test should be performed on the site. If necessary, a sample should be sent for testing the maximum expansion of the foam.

3.2 Field Determination of Density[41]

A simplified but equally precise test is used to determine core density during a field fabrication. The density is measured by means of a buoyancy test.

Principle: After determination of the mass of the specimen, the volume shall be determined by water displacement.

Apparatus: A balance, readable to 0.01 gram

                   A knife with minimum 180 mm (6 in.) blade

                   A graduated cylinder, minimum: 1000-ml size with 0-ml graduation

Procedure: Cut and remove three pieces, each approximately 150 mm- (6 in.) long and 50 mm- (2 in.) wide through the whole thickness of the foam in the middle of the polyethylene area. Then, perform the following operations:

1. Without weighing the pieces, make sure that they would fit into the graduated cylinder, trim when necessary.
2. Weigh the specimen and record its mass.
3. Fill the graduated cylinder with the required volume of water and record the volume.
4. Submerge the foam completely using a thin profile, e.g., a thin-wall plastic tube, and record a new volume of water. (The test must be done within one minute.)
5. Calculate the water displaced by subtracting the volume recorded in (3) from that in (4).
6. Divide the specimen mass in grams (2) by the water displacement in liters (5) to obtain a density in g/L that is identical to that in kg/m³.

3.3 Maximum Temperature and Time of Its Occurrence

The maximum temperature rise test should be performed if spraying of the test pattern gives inconclusive results.

Often, this test is performed on the back side of the 600 mm × 600 mm (2 ft × 2 ft) clean, dry, plywood sheet, which is later used as the external test sample.

3.4 Adhesion and Cohesion Strength

Principle of the adhesion test: The foam is cut through the whole thickness of the layer to provide place for an adhered circular plywood disc equipped with a hook. This disc is bonded with epoxy (or another adhesive) to the SPF sur-

---

[41]See: National Standard of Canada 51.39-92.

face. A specified load is attached to the hook perpendicular to the surface. The load may be increased until the rupture is observed.

Apparatus: Two round plywood discs with diameter of 70 mm and thickness of 20 mm

> Two component epoxy adhesives
> A core tool made from thin–wall tempered pipe to make cylindrical cuts with 71-mm inside diameter
> Two wires with hooks
> A support frame to provide a 90° angle for the wire loaded with the weight (e.g., pull-out machine similar to that used for single ply testing)
> A series of calibrated weights

Procedure: Select two areas in the middle of the plywood substrate and make cylindrical cuts throughout the entire thickness of the foam layer. Bond the discs. Pass each wire through the support and attach the minimum weight of 1 kilogram (slowly to avoid impact loading). Observe if any delamination occurred. Increase the load stepwise and observe again.

Cohesion test. The test is performed in the same manner. The bore cut, however, is made to approximately half of the insulation thickness.

This test may either be performed on the standard plywood sheet, or used to verify the substrate conditions in re-covering the old roof when adhesion to the substrate is in question.

Note: a sample should be prepared for external testing if either

- the results of adhesion/cohesion strength test fall below limits established during the initial product characterization
- the maximum temperature indicates a significant difference between the initial foam characterization and the actual field application

3.5 The Sample for External Testing

This sample is sprayed on both sides of a 600 mm × 600 mm (2 ft × 2 ft) clean, dry, plywood sheet in the manner specified in the chapter dealing with testing. One side of the plywood sheet is half-covered with polyethylene to facilitate easy removal of the foam for density and compressive strength testing. The exposed plywood side may be used for adhesion/cohesion tests.

Typically, the external testing would include adhesion to the substrate, tensile and compressive strength, and foam dimensional stability. These tests can be done either by the manufacturer or by an independent testing laboratory.[42]

---

[42]Inquire at the SPF Division of the SPI Inc. or at CUFCA as to the testing laboratories having national accreditation and sufficient skills in the SPF testing area.

CHAPTER 7

Implementation of the
Performance–Based
Quality Management

THE SPF CONTRACTOR fabricates the SPF product on site in accordance with manufacturer's instructions. Nevertheless, the contractor may select one of many different foam systems, each exhibiting different thermal, mechanical, and fire performance characteristics. Therefore, the SPF contractor must understand different aspects of the construction process and not only be trained in the art of polyurethane foaming.

The outcome of the contractor's work will be affected by several installation parameters such as thickness of the foam, number of passes, and air and substrate temperature.[43] While the conditions of installation have a pronounced effect on all properties of the installed SPF layer, its performance is also affected by the design of the system itself. In short, performance of the building envelope component with SPF depends on understanding materials and their interaction with the environment by all people involved in the design, selection, fabrication, and maintenance of SPF systems.

Furthermore, the simple act of SPF fabricating at a location, is an effect of a complex process that involves five different industrial groups, each more or less willingly contributing to the final performance of field fabricated SPF insulation:

1. Manufacturers of raw materials such as polyisocyanate (component A), polyol, blowing agent, catalyst, surfactant, flame retardant and other additives (antioxidants, fillers) that are used to modify properties of the foam (component B)[44]

---

[43]Normally, SPF should not be applied at temperatures lower than 40°F (5°C), although specially designed foam can be applied even at 25°F (–5°C). Normally, SPF should not be applied when air relative humidity is above 70% RH and wind exceeds 15 mph (7 m/s), unless wind screens are used. Furthermore, if the difference in substrate and air temperature exceeds 8°F (5°C) an adjustment in temperature of the proportioning unit (sometimes also the gun pressure) is needed.

[44]There are approximately sixteen major suppliers of chemicals and blowing agents.

2. System developers, who blend and formulate polyol mixtures for a specific application[45]
3. Manufacturers of spray equipment[46]
4. Manufacturers of coating and protective finish systems
5. Polyurethane foam contractors, who purchase an SPF system manufactured by (2) and use equipment of (3) to fabricate the foam with a protective covering manufactured by (4)

One may use a spray, pour or froth polyurethane system (depending on whether the reaction starts before or after the foam application). Even within the spray systems, the reaction starts in such a broad range as 2 to 16 seconds. The fast foams are used in wall applications, and the slower are used for roof applications. HD-type and SHD-type SPF with densities above 45 kg/m³ (2.8 lb/ft³) are used in roofs. MD-type and LD-type SPF with a density below 37 kg/m³ (2.3 lb/ft³) are used in wall applications. Closed-cell foams with very low densities have been successfully used for air sealing applications.

Historically, improvement of materials and application systems had taken place over a long period in the form of small incremental changes. Introduction of new blowing agents, stabilized bromine compounds as fire retardants, wider ranges of application temperatures, new polymeric systems, fast-drying coatings, and improved application machinery (foaming robots) may illustrate the progress in this field. The outlook for SPF appears very good. Tye (1987) concluded that SPF has desirable characteristics such as:

> the production of monolithic joint free layers, potentially simple installation, consideration as both air and/or moisture retarders for appropriate applications, and the ability to choose from a range of formulations which provide suitable physical, mechanical and thermal properties for a particular part of the envelope. Significant growth potential is foreseen "where reduction of the effects of thermal bridges and air movement can be achieved."

The same assessment stated: "Adequate information on long-term performance and specifically on some relevant parameters which affect final thermal performance of materials, coverings and their combination does not exist. This includes coefficients of expansion, dimensional stability, water absorption, air and moisture (vapor) permeance, adhesion and cohesion especially at different temperatures."

This monograph, primarily based on knowledge developed during seven years of the joint SPI/NRC projects, answered many of the questions asked by Tye (1987). Methodology to predict long-term thermal performance of

---

[45]There are fifteen to twenty system houses in North America.
[46]There are four major manufacturers of equipment used by SPF contractors in North America.

SPF has been developed and verified. Long-term thermal and mechanical performance of SPF manufactured with different blowing agents was carefully evaluated. Study of thermally driven moisture (movement caused by thermal gradients) created basic understanding of moisture transport and storage (absorption) in SPF. These laboratory investigations were designed to explain many of the findings from long-term outdoor exposures, such as studies of SPF roofing systems performed by Alumbaugh or Kashiwagi. The improved understanding provides the SPF industry with new opportunities and this industry is well-positioned to meet the challenges of the coming decade.

While there are many new opportunities for the SPF industry, the extent to which these opportunities may be realized depends on the industry determination to maintain the leading system of quality assurance. The research reviewed in this monograph has shown that HCFC-blown sprayed polyurethane foam may have long-term mechanical and thermal performance as good as or better than the CFC-11 blown foam. Whether consumers (specifiers) will accept that a new generation of SPF is better than the old one, depends primarily on Quality Assurance programs.

To promote and expand qualified use of SPF in North America, two organizations—the Spray Polyurethane Division of the Society of the Plastics Industry Inc. and the Canadian Urethane Foam Contractors Association Inc.—have established extensive quality assurance programs. These programs include the classroom training and field training that together with qualified experience, professional networking, and updated material from different technical and industry promotion committees of the SPFD[47] constitute a strong basis for quality in the SPF industry.

The SPF industry wants to make sure that the SPF products provided to the consumer are of the highest quality, quality equal to or surpassing that of plant-manufactured materials. A field-fabricated product such as an SPF foam can offer technical benefits to the consumer. Industry-wide quality assurance takes away the "uncertainty" of field fabrication.

These programs may differ with local economic conditions and requirements of building officials, yet, they have the same technical basis over the whole North American continent. These programs include installer training and certification through rigorous accreditation courses and apprentice training and examinations for certified installers. The installer certification is a prerequisite for licensing the contractors on a yearly basis and providing them with an ID card.

---

[47]A close technical collaboration that exists between SPFD/SPI and CUFCA ensures the same technical standards are applicable over the whole North America.

In some areas, CUFCA has established a QA program, which provides an unannounced inspection system with spot surveillance (a record of each job site is forwarded to CUFCA to build a database tracking system).

The industry program includes a multi-level training curriculum combined with actual field experience to produce the required level of competence. There are four levels of training, with about one year of working experience ascribed to each of them, but any installer with a proven two-year practice is allowed to complete the course and take the examination. Typically, the program includes:

1. Introduction to foam systems, safety, and transportation issues (apprentice level)
2. Equipment introduction to foam applications, properties of SPF, troubleshooting
3. Product knowledge and advanced foam applications
4. Building science, client relations, managing the job, field testing

The SPFD accreditation programs are organized similarly, on basic and advanced levels, though they may differ in details. They are exemplified in Tables 37 and 38. Table 37 specifies information modules involved in training for different positions within the contractor's business. Table 38 gives a more detailed example, namely, the first course on fundamentals of SPF and coating systems.

In Canada, the self-control of the SPF industry has also been supported by requirements of building codes and provincial regulations. For instance, the National Standard of Canada, ULC S-705 requires that the SPF contractor's training must include the following:

- description of chemical components, including their property and MSDS for each component (to be provided by the manufacturer)
- safe handling and use of chemical components (also provided by the manufacturer)

Table 37. Contractor training as organized by the SPFD/SPI Inc.

Applicator (Level A)	Foreman (Level B)	Sales (Level C)	Management (Level D)
Fundamentals	Fundamentals	Fundamentals	Fundamentals
Roofing	Roofing	Project control	Roofing
Equipment	Project control	Professional selling	Project control
Materials	Equipment	Roofing	Professional selling
Roof inspection	Materials		Finance & accounting
	Roof inspection		Managerial skills

Table 38. Course 101 on fundamentals
of SPF and coating systems.

Module	Content
1	Introduction
2	Industry standards and coating specs
3	SPF chemistry
4	SPF equipment
5	Foam applications and workmanship
6	Coating chemistry, properties, and safety
7	Coating equipment
8	Coating applications and workmanship
9	Job site quality control
10	Inspection, maintenance, and repair
11	SPF self-defense/safety

- operating parameters and maintenance of the equipment (equipment manufacturer)
- physical properties of the SPF and limitations caused by the conditions of installation and use of the foam[48]
- site and substrate preparation
- chemical storage and handling, including disposal of waste material
- limitations for use of SPF insulation
- applicable codes and regulations

This monograph brings these two industrial accreditation and licensing programs even further by introducing concepts of the performance-based quality assurance which combines technical specifications, performance assessment and on-site testing through the PBQA system. In this manner, the SPF industry can provide bidding documents that characterize long-term performance of SPF in construction systems and predict energy use for many future years.

---

[48]This information is to be developed by the system manufacturer.

PART TWO

The Systems Approach to Buildings

CHAPTER 8

Principles of
Environmental Control
in Building Envelopes

A S STATED IN the preface, this monograph also addresses some "theoretical" aspects of building science such as principles of design for durability and long-term thermal and moisture performance of materials.

1. INTRODUCTION

In the past, building envelopes were leaky, and natural ventilation was relied upon to bring fresh air into buildings. Until the energy crisis of the late 1970s, energy was inexpensive. Our approach to environmental control of buildings has changed since that time.

Increased tightness of building envelopes, controlled ventilation and air conditioning systems are relatively recent changes in the building technology. Recent trends to thicker thermal insulation in cold climates caused the primary exhaust device, the chimney flue, to be used less frequently.[49] The trend towards using electric heating, heat pumps, and power-vented sealed combustion furnaces has further eroded the role of a traditional active chimney. Chimney flues acted as exhaust fans, which extracted great quantities of air from the conditioned space, reducing the moisture load acting on the building envelope. Less efficient chimney flues increased the indoor air humidity, increasing frequency of vapor condensation on surfaces of windows or thermal bridges. More recently, the significance of this issue was reinforced by increased epidemiological evidence that links respiratory diseases to dampness (Robinson, 1992) and growth of specific mold types.[50]

---

[49]In the extreme, reducing a chimney's ability to exhaust products of combustion may lead to spillage of combustion products, backdrafting of furnaces and fireplaces, and associated health and safety problems.
[50]Jim White, panel presentation at 7th Conference on Building Science and Technology, Toronto, Canada, March, 1997.

1.1 Indoor Environment

Building enclosures have become significantly tighter, reducing the exchange of air between the indoor and outdoor environments. The lower the air change, the less effective the dilution of pollutants in the indoor space. New consumer products increased the variety of pollutants in the indoor air. These pollutants include moisture (from people and appliances), formaldehyde (from particleboard and furnishings), volatile organic compounds (from carpets, paints, cleaners, and adhesives), radon (from basements, crawl spaces, and water supplies), and carbon dioxide (from people). The reduced dilution caused concentrations of these pollutants to increase even further.

In some residences, the installation and use of indoor spas, hot tubs and central humidifiers made moisture control more difficult. In commercial and institutional buildings, the installation of pressurized and humidified computer rooms and special use areas such as health clubs or copying/duplication rooms created additional environments that are hostile to humans.

These changes in construction processes introduced new considerations for materials.

1.2 Material Considerations

Any material must be assessed in the context of a system. What is the function of the material? Can the material perform the function? What is the risk to the occupants, the building, and to the environment? Extending this argument further, there are no truly benign materials, and nothing is completely risk free. However, risk can be managed. For example, a toxic material can provide significant benefits and pose little risk when used properly. Use of some dampproofings on the exterior of a preserved wood foundation illustrates this point. An inherently toxic material provides substantial benefits to the system (moisture control) and does not pose a high risk to the occupants.

Concerns have been raised that many synthetic agents impact the indoor environment and that "natural" or "green" materials should be used instead of them. However, many "natural" materials contain volatile organic compounds, are potent irritants, and pose hazards to health. Allergic reactions to the odors emanating from "cedar" closets and chests are common. As well, radium and radon are natural materials. Yet, nobody would suggest that these two "natural" materials be used in construction (radon could be used instead of argon in sealed glazing units; radium could make thermostat dials easy to read at night).

The risk to occupants is low if a particular agent remains in the building product and does not affect people through respiration and physical contact.

In general, building materials, which do not off-gas are preferable to those that do. Products that off-gas a little[51] are preferable to those which off-gas a great deal. Less toxic alternatives should be used in place of alternatives that are more toxic. The principle of product substitution should be employed wherever possible.

For volatile organic compounds (VOC), the decay of their emission rate must be considered. For example, most interior paints contain significant quantities of VOC (solvents). They off-gas or evaporate very rapidly and leave behind a relatively "benign" surface. After some period these materials pose little risk to most occupants over the rest of the service life of a building. However, they may pose a real "risk" to the painters as an occupational hazard.

Superimposed over all of the specific concerns about material and product use on the interior environment of a building and the occupants, are the concerns relating to the local and global environments. Is it more appropriate to use a recycled material in place of a new material? Is the new material or product manufactured in a non–disruptive or least-disruptive manner to the environment? How much energy was used to make it?

These are all valid and important concerns. Sometimes, however, when pursuing one, small concern we lose sight of the bigger picture. For instance, a large embodied energy content is often voiced against the use of foam plastic insulation. Franklin Associates (1991) calculated energy consumption through each stage of a product's life cycle. This calculation included the energy input involved in acquisition of raw materials, materials manufacture, material waste, product and by-products manufacture, production waste, cost of transportation to the first buyer, and a number of other variables. It was established that 50 Btu's are used to manufacture 1 pound of polystyrene insulation.

Compare the embodied energy with energy saved over a period of thirty years (as an example of the service life, or a mortgage period). With density of 1 lb/ft³ (16 kg/m³) and thickness of 2 inches (50 mm), one pound of insulation covers 6 square feet (0.04 m²). For a seasonal difference of 32°F (20°C) over a 5-month heating period, the energy saving achieved by an insulation with thermal resistivity of 3.5 (ft²hr°F)/ (Btu in) is 3 millions of Btu (30 years × 5 months × 30 days × 24 hours × 32 degree F × 6 square feet / 3.5 × 2 in = 3 millions of Btu). It is evident that any type of thermal insulation is a "green product" in terms of the saved energy. The embodied energy should not be a factor affecting the selection of thermal insulation.

As highlighted above, to design a durable and cost-effective building shell, we must select materials with a view to their contribution to the subsystem.

---

[51] A typical MD-type SPF was shown in Chapter 2, Section 5 to have the half-life period for HCFC off-gasing equal to 181 years.

Evidently, the same approach should be used when selecting subsystems.

1.3 Subsystem Considerations

Discussing the architectural interior systems, Flynn and Segil (1970) listed four subsystems:

1. The site subsystems (wind diversion, sun shading etc.) that provide an environmental context of the building (often called the mezzoclimate)
2. The building envelope
3. The subsystems that satisfy environmental and service demands
4. The subsystems that facilitate the distribution of energy to and throughout the buildings which together (3) form a comprehensive environmental system

Flynn and Segil (1970) stated: "But rather than a simple correction of climatic deficiencies, the environmental control function of building must be oriented toward the more extensive sensory demands of various occupant activities and experiences. This occupant perceives light as the surface brightness and color; he absorbs heat from warmer surfaces and warmer air; and he himself emits heat to the cooler surfaces and cooler air. He responds physiologically to humidity, to air motion, to radiation and to air *freshness*. He also responds to sound. A major function of the building, then, is to provide for all the sensory responses concurrently—to establish and maintain order and harmony in the sensory environment."

Is this often happening in practice? No. Today's process of design, construction, and commissioning appear to accommodate the separate roles of construction professionals and building trades (each having the circle of responsibility corresponding to their expertise). Often, this system does not ensure a team communication with a view to satisfying the user's requirements. For instance, testing professionals specify very precise and detailed procedures, which ensure constancy of material performance. Any piece of material that passes the test criterion is as good as one that was originally tested. The result of the test itself, however, seldom describes the field performance of this material in a construction system (Bomberg, 1982). Only when enough data from field performance has been collected and correlated with the test, can such a test have a judgment value for a designer selecting materials.

The code professionals write requirements that carefully describe traditionally proven solutions. Architects, structural engineers (who design the building shell) and mechanical engineers (who design the HVAC and services), and fire, acoustic, lighting, and material experts are carefully trained in their respective professions. Architects and civil engineers receive no training in the building science. In effect, there is no common ground for understanding "how to establish and maintain order and harmony in the sensory environment" when developing design for the "durable and cost-effective building shell."

The current design process does not provide the facility for predicting performance of the new system during the design stage. Neither does it provide means for an effective quality assurance that starts during the design and continues through construction ending with the commissioning of the completed building.

2. FACTORS AFFECTING THE DESIGN PROCESS

Before one can improve a process, one must understand it. Therefore, our analysis must start at the beginning of the design process.

2.1 User Requirements—The Starting Point of Design

Primarily, the building envelope provides a shelter from the outdoor environment to enclose a comfortable indoor space (Hutcheon, 1953). In doing so, the envelope must withstand many mechanical and environmental forces over its service life. These forces include climatic factors such as temperature, air, and moisture in their various forms. In climatic extremes, for instance the Canadian cold, the envelope must be well insulated to provide the required level of thermal comfort. Other comfort considerations involve noise and fire. These functions must be achieved at a reasonable cost.

The user requirements listed in the first column of Table 39 are general. They must be, therefore, formulated as specific, measurable[52] requirements for which evaluation procedures may be developed and acceptance criteria established (Bomberg, 1982). Table 39 highlights that one user requirement may result in a multitude of technical considerations, which may or may not be expressed as performance characteristics. These requirements may be needed to define attributes of the factor that serve to attain a specified construction objective. To incorporate both the performance and the descriptive requirements that follow out of the user needs, we will use the term "performance objectives."

Performance objectives listed in Table 39 differentiate between time-independent and time-dependent effects. The latter are often called "durability" if the process causes a damage to the material, and "serviceability" if the process reduces the level of their performance in the building envelope.

---

[52]Measurement is the first step that leads to control and eventually improvement. If you can't measure something, you can't understand it. If you can't understand it, you can't control it. If you can't control it, you can't improve it.

Table 40 shows that damage caused by the time-dependent failure mechanisms prevails over other types of damage. It shows that time-dependent aspects account for 60 percent of the observed damage (Gertis, 1982). Similar observations may be found in the UK study by Harrison (1983). The durability considerations are the most difficult part of building envelope design and evaluation as they involve many different and interacting variables and as they depend on environmental and service conditions as well as period of service.

Incidentally, a popular concept of materials being either durable or not is false. The durability of a material in a building envelope depends on the outdoor and indoor climate, type of construction, and conditions of service. A small change in one of these variables may result in material failure during the first year or a flawless performance for forty years. For instance, a clay-brick veneer enclosing unheated storage space showed a continuous spalling, while the walls enclosing the heated space of the same building were undamaged.

Judging from the types of failures listed in Table 40, it is evident that the designer needs more guidance on how to design durable building envelopes.

Table 39. *User requirements versus performance objectives for exterior walls.*

User Requirements	Performance Objectives Time Independent	Performance Objectives Time Dependent
Space separation	Strength and rigidity (deflection): *utility loads, wind, impact loads*	Weathering and aging: *time-dependent property changes*
	Relative movements and dimensional changes of materials	Fatigue, deformation: *sealants, gaskets*
	Vibration and noise: *airborne noise and structural vibration*	
	Aesthetic considerations: *color and texture of the façade*	Weathering: *stain, discoloration*
	Risk and prevention of fire: *combustibility, smoke development, toxicity, time for escape*	
Environmental control	Control of heat flow: *heat gains and losses, thermal bridges*	Thermal deformation and stress
	Control of water: *rain, ground water, condensation*	Hygric deformation and stress moisture accumulation leading to deterioration: *corrosion, freeze-thaw, rot, efflorescence, mold, mildew, weathering, etc.*
	Control of air flow: *wind, stack effect, HVAC operation*	
	Control of water vapor: *air flow, diffusion and thermally driven*	
Cost	Initial cost, maintenance, and repair	Life cycle cost

Table 40. Frequency of damage in German
wall systems (Gertis, 1982).

Cause of Damage	Percent
Temperature	13
Restrained movement	12
Moisture	9
Deformations	9
Settlements	7
Creep	6
External climate	4
Total of the above	60
Other causes	40

2.2 Predictability of Building Performance—A Historic Background

The relation between knowledge and predictability of the construction performance was examined by Hutcheon (1971). He observed that:

> The knowledge about building, called, for convenience, *Building Science*, is valuable largely because it is useful in predicting the outcome or the result of some building situation. The situation may be real, if the building already exists, or may be posed in a hypothetical way in the normal course of building design. Rational design is possible only when there is the capability to establish, each time a choice is made, the probability of a particular result.

Does the construction experience based on tradition promote rational design? The answer is partly affirmative:

> For tradition embodies prediction, embracing those things which have been shown by experience to produce a predictable result. Such experience very often has arisen from unintended, costly, full-scale experiments associated with failure of part or all of a building during or after construction.

There is a clear limit of the use of tradition: "Tradition has a great weakness in that it deals only with a way of doing something, without any contribution to understanding of why the traditional method works. This being so, it is usually not possible to identify the important factors either in the situation being served or in the arrangement or solution provided." There is the crux of the matter, and Hutcheon continues. "The experiments must be done if predictability is to be extended. They can be done more economically, and with greater return, if devised and carried out in a systematic series, which is, of course, research. They may be done in the laboratory as well as the field, often on model scale."

Would a laboratory program provide a good means for achieving predictability of performance? Again, the answer is partly affirmative. One must remember that the knowledge (the building science) was defined as a synthesis of the understanding and the experience. The advantages or disadvantages of testing must be analyzed in this context.

With knowledge about similar situations, one can identify factors that influence the results and select tests providing the needed information. When little is known, an elaborate test program may be needed. Cost may limit the research program to be undertaken, posing the risk that "in the absence of knowledge the choices made may fail to represent the service conditions in some important way. The single test, by itself, provides very limited information; it can be very effective when designed to provide some critical information in an otherwise adequate body of knowledge, like the final piece of a puzzle. Its value depends almost entirely on the relevant knowledge already available" (Hutcheon, 1971).

This discussion outlines *a paradox of knowledge and testing.* To design a simple and effective test a large body of knowledge is required; to develop such a body of knowledge, a large number of carefully planned tests are needed. Addressing this issue, Bomberg (1982) postulated development of a series of interdependent ASTM tests, so-called blocks of test methods.

A particular difficulty relates to predicting long-term performance of new products. A manufacturer has little control over application of the building product. Furthermore, evaluating the product durability may involve the outcome of several interactive and cumulative degrading effects. Hutcheon highlighted the limitations for such predictions:

> This is exceedingly difficult for new materials and must be based entirely on the knowledge of the product and of related products and situations until supporting evidence from significantly longer periods of use becomes available. There is always great demand for accelerated durability tests, and these are very difficult to devise and to verify. Final verification must await the completion of a lifetime of service.

This discussion outlines *a paradox of durability evaluation.* There are no methods of accelerating weathering and aging processes, there are methods involving either more severe exposure conditions or the theory of mechanical similitude. Extreme environmental conditions (e.g., elevated temperature) may accelerate some physical or chemical processes. Some methods may reduce the period of testing by altering properties of material or scaling geometrical relations. As an example, a transport process may be accelerated by changing the ratio of exposed material surface to its volume as in the scaling factors to accelerate aging of foams (Isberg, 1988; Sandberg, 1990; Bomberg, 1990; Christian et al., 1991).

To relate the results from the so-called "accelerated test" to performance in the actual conditions one must know the extent of "acceleration" obtained in the laboratory (differences in the severity of the exposure with regard to all factors affecting the final outcome).

It is evident that to predict long-term performance of a construction system one needs such a scientific basis that would allow one "to assess the relevance of experience and thus to draw upon broader and more varied experience in the development of predictability" (Hutcheon, 1971).

The research aiming at enhanced understanding of general functional relations has dominated building science during the 1970s. The performance analysis (Wright, 1972; Blach and Christensen, 1976; Cullen and Sneck, 1980; Becker, 1985), was thought to be a panaceum for enhancing predictability of performance for any system of the designer's choice. Despite concentrated efforts of many international groups, the performance analysis has failed to become a part of construction practice. Why did it fail? Perhaps, because it lacked a mechanism to combine the holistic and analytical approaches and could not produce a synergy between the two pillars of building science: engineering experience and understanding of scientific principles. Perhaps, because it did not recognize the dual nature in the process of design and evaluation.

2.3 Addressing the Duality of the Design Process

Designing for environmental control of the building envelope assembly compels professionals to integrate two very different conceptual processes. One extreme encompasses analytic thinking, involving testing and calculations; the other encompasses analog (lateral) thinking, based on broad experience and judgment based on an understanding of what makes a building envelope function. On the analytical side is a complex array of tools, models, and data, that describe the material, structural, and environmental factors relating to the building envelope. On the qualitative side is a sense of how a particular building envelope would function.

The design of an air barrier system offers an example of how the process of dual track and iterative design might work. The information flow may start with a search for suitable materials. Typical questions are asked about possible materials and their air permeability, ability to be extended, pliability, adhesion, means of attachment, connection, and support. The review would also address the long-term aspects of performance (material weathering and aging), stress and deformations during service, and projected costs of repairs and maintenance.

After making an initial selection, the designer then specifies the details such as intersections and joints between building elements (for example, foundations, walls, floors, windows, and doors). Then, to achieve satisfactory performance in these locations, the designer must ask further questions concerning the performance of the whole system, such as probable location of air leakage, rate of air leakage, its impact on vapor condensation, and possible damage. Throughout the design process, the designer consults with structural, electrical, and mechanical experts to ensure that the selected materials will perform satisfactorily.

In addition, the designer reviews the buildability aspects such as material installation under different weather conditions, level of labor skill required for installation, accessibility to perform sequence of tasks and expected level of construction tolerance. Buildability, as the word suggests, reflects whether the design created on paper can be constructed with the resources available.

Even with the best design, the probability that no defects will develop during the service life of the building envelope is so low that some redundancy in the design becomes necessary. For instance, the plane of the air barrier system may be incidentally punctured, or poorly connected to an element of the assembly, for instance, windows. The designer must then evaluate whether this excessive moisture could be drained or dried out. How long would the drying process take and what effect may it have on other materials? Are these materials sensitive to moisture? Can the prolonged presence of moisture cause corrosion, differential movements, mold growth, or rot?

Because many qualitative decisions (though based on experience) appear arbitrary, some people attempt to replace them with more "stringent," analytic evaluation criteria. Consider the example of a vapor barrier.[53] A typical vapor barrier is required to have a permeance of no more than one perm, a unit that represents sufficient retardation of water vapor flow for traditional wood frame housing. For most building authorities using a layer with 1.5 perms appears out of the question. Yet, calculations made with a complex model of heat, air, and moisture transport demonstrated that permeance values ranging from 0.2 to 7 perms are suitable for various combinations of materials and climatic conditions in Canada (Karagiosis and Kumaran, 1993). Ojanen and Kumaran (1996) showed that with an air barrier system in place, a vapor barrier with permeance of 3 perms would satisfy most locations in Canada. The latter requirement can easily be satisfied by primed and twice-painted gypsum board (without vapor barrier paints).

3. DEFINING THE ENVIRONMENTAL CONTROL

The design process should occur simultaneously on different levels. The building should be analyzed as a whole at the same time when each of its components is analyzed. The environmental control must also be a part of this

---

[53]Changing the name to vapor retarder may appear scientific but does not address the issue.

analysis if the interactions and trade-off between control of heat, air, and moisture transports are to be fully realized.

3.1 Interactions of Heat, Air, and Moisture Transports

Heat, air, and moisture transports across a building envelope are inseparable phenomena. Each influences the other. All transports, however, are influenced by materials contained within the building envelope. Often we simplify the design process by ascribing control of each phenomenon to a particular material. The thermal insulation is to control heat transfer and the air barriers to control air leakage. Likewise, to eliminate ingress of moisture, we use the rain screen and the vapor barrier.

However, each of these materials may perform different functions and influence several aspects of the overall performance. For instance, by controlling air leakage, the air barrier provides an effective moisture control. Similarly, by increasing temperature in the wall cavity in cold climate, an external insulating sheathing reduces the intensity of vapor condensation in the cavity of a frame wall.

To ensure that all aspects of the building envelope perform effectively, we must deal with heat, air, and moisture transport collectively. In some ways, this approach represents a return to the thinking of sixty years ago, long before detailed performance analyses became routine. Today we have improved standards and requirements concerning performance of the individual components of the building envelope. Perhaps, we are better able to apply the fundamental concepts first introduced in the 1930s.

3.1.1 AIR TRANSPORT—A LESSON FROM HISTORY

Air transport represents a critical factor in environmental control. It underscores virtually all facets of environmental control as it moves heat, moisture, pollutants or contaminants, and smoke if there is a fire situation. (Air is necessary to maintain combustion and is therefore critical for the spread of fire.) Flows of heat, air, and moisture critically impact durability of the building envelope.

Our understanding of the performance of walls comes mainly from cold climate and primarily from the Prairie regions of North America where the climatic extremes magnify any faults in the ability of building envelopes to maintain environmental control. Research on air leakage through frame walls, performed in the early 1930s, led to the acceptance of building paper. The building paper reduced heat losses by limiting the passage of air, and improved indoor comfort by reducing drafts, while it permitted water vapor to pass to the outdoors. The building paper even reduced moisture damage to the walls by preventing wind washing (wind entering and leaving to the outside) which decreases the inner surface temperature. Use of the building paper as the weather barrier that permitted the wall to breathe became an important part of wall design.

At the same time, in the quest of thermal comfort, wall cavities became filled with insulation—first wood chips stabilized with lime, then shredded newsprint and eventually mineral fiber batts. Although water vapor passed through this thermal insulation as easily as through the air layer, the presence of insulation reduced temperature on the exterior part of the wall cavity causing interstitial condensation of water vapor.

A vapor barrier was then introduced on the warm side of the wall cavity to reduce ingress of vapor from the warm space. Consequently, the walls of homes built in Prairie regions in the 1940s already included the outside weather barrier and the inside vapor barrier. Although materials used in these barriers were based on building paper with or without asphalt impregnation, environmental controls in building envelopes were developed more than fifty years ago.

3.1.2 THERMAL PERFORMANCE—ENERGY AND HEALTH

This assessment involves three types of considerations:

- quantity of heat transferred through the walls, windows, and other elements of the building envelope
- quantity of heat needed to bring the temperature of the outdoor air to that of the indoor air—the air leakage (or rate of air exchange) component of energy
- differences in temperatures on the inner surface of the building envelope—the risk for mold growth

The heat transfer through the walls may be described with six levels of accuracy:

1. Considering only the insulated area of the wall under the steady state conditions.
2. Considering only the unidirectional heat flow under the steady state conditions (parallel path model through the clear wall area).
3. Considering multidirectional heat flow (clear wall area) under the steady state conditions.
4. Considering multi-directional heat flow-through the whole system including corners, junctions, etc., under the steady state conditions.
5. Considering multi-directional heat flow-through the whole system including corners, junctions, etc., under the transient heat flow conditions (no input from air and moisture flows).

6. Considering multi-directional heat flow-through the whole system including corners, junctions, etc., under the transient heat, air and moisture flow conditions. The transient airflow conditions are defined by multizonal air flows and interaction with HVAC equipment. The transient moisture conditions involve wetting and drying (rain, condensed moisture, etc.).

The first approximation considers only the plain, insulated areas of the wall. For instance, a frame wall insulated with RSI 3.5 glass fiber batts is called an RSI 3.5 wall.

The second level of accuracy is not much better. An actual thermal resistance for each section is used; however, the model assumes no deviation of heat flow path through the wall, i.e., strictly unidirectional heat flow. With the area of thermal bridge[54] typically 2 to 3 percent, the increase in the overall heat transfer is very limited.

The third level of accuracy incorporates effects of multidirectional heat flows caused by thermal bridges. Kosny and Desjarlais (1994) define a clear wall area as the part of the wall system free of thermal anomalies such as corners, window and door openings or joints with other structural elements. In the discussed case, because of the wood framing, the RSI 3.5 (R20) wall becomes an RSI 3.1 (R17.6) wall.

Historically, the shift from extensive measurements of heat transmission through the structures (Pratt, 1969) to computer calculations (Kosny, 1995) was slow. In the mid 1980s a discussion took place whether measuring or calculating thermal performance of the structures is preferred (Wagner et al., 1984). Calculations overestimated wall systems comprising air spaces (Greason, 1983) and reflective insulations (Hollingworth, 1983). Calculations, however, because they permitted both steady state and transient conditions to be analyzed (Kuehn and Maldonado, 1984), replaced most[55] full scale, steady state thermal measurements.

The fourth level of accuracy adds effects of other thermal anomalies such as wall corners, and wall-floor junctions, while assuming that the steady-state representation sufficiently describes the thermal performance of the building. Kosny and Desjarlais (1994) performed 3-D calculations for a 1-story ranch house. The overall thermal transmittance of the wall falls in three classes:

- about the same or smaller than for the clear wall (EPS-forms and Larsen-truss walls)

---

[54]Materials or elements having much higher thermal transmittance than the typical cross-section of the wall called thermal bridges.
[55]Light gauge steel frame walls still need an experimental check.

- about 10 percent larger than that of a clear wall (wood stud and skin panel wall systems)
- about 20 percent larger than that of a clear wall (steel frame wall)

Knappen and Standaert (1985) presented comparisons between one-, two-, and three-dimensional solutions for a double-glazed wall (A3) insulated with 6 cm cavity insulation. The error of the 1-D estimate was 34 percent and the error of the 2-D estimate was 16 percent. A similar magnitude of multidimensional effects was reported from field studies. For instance, Fang et al. (1984) reported that the effect of thermal bridges increased thermal transmission by 10, 11, and 21 percent when compared to walls without thermal bridges.

The fifth level of accuracy in the assessment of thermal transmittance deals with the transient weather conditions and thereby induces the effects of thermal mass on heat loss from, or gains to, the indoor space. Normally, such a calculation is performed for a specific climate and construction type with a recognized computer model, e.g., DOE 2. The effect of building mass on annual heating energy requirements is normally of much smaller benefit in cold and extremely cold climates (Mitalas, 1979) than it is in mixed climates (European Passive Solar, 1986).

The sixth level of accuracy involves effects of moisture. It is used only for durability assessment where coupling between thermal and moisture gradients may significantly affect distribution of moisture in the materials leading to loss of structural integrity or long-term performance.

Brown and Stephenson (1993) measured heat transmission characteristics of walls under transient (dynamic) conditions to confirm the data and procedures provided by the ASHRAE Handbook—Fundamentals. For all specimens, the measured and predicted frequency response agreed well. "On the other hand, the measured thermal resistance varied 45% to 90% of the predicted thermal resistance," concluded the authors. (The fact that results obtained from heat transfer models may or may not deviate from the measured values depends on a number of approximations in material properties and boundary conditions.)

The above discussion showed that the accuracy in determination of thermal transmittance can vary by an order of magnitude. What method should the designer use?

The designer may use two different methods. In the conceptual stage of design, a simple concept of thermal resistance (R-value) is absolutely sufficient. (Though these concepts were developed for comparative purposes, they may fail to describe thermal performance of some systems (e.g., slab on ground) or be imprecise for other systems.)

A recommended approach for the conceptual design stage is to use a general equation:

$$R = (R_1 + mR_2) / (1 + m) \tag{1}$$

where: R_1 is the thermal resistance calculated from the parallel path model, and R_2 is the thermal resistance calculated from the isothermal planes model.

The unidirectional heat flow model was introduced in this paper as the second stage of accuracy in R-value measurements. The isothermal plane model, which assumes a perfect equalization of temperature within each parallel wall section, is another limiting case. Equation (1) states that the thermal resistance of a wall is restrained by these limits.

There was a substantial amount of work related either directly or indirectly to the m-factor in Equation (1). A 1960s Russian thesis analyzed different insulated and hollow blocks showing the value of m between 0.1 to 80 with the most probable $m = 2$. Therefore, if the difference between R_1 and R_2 is limited, some European standards used $m = 2$ (Plonski, 1965). Brown and Schwartz (1987) showed that $m = 1$ approximated most of the insulated wood frame walls. Garrett (1979) showed that the relation R_2/R_1 varied between 1 and 1.7 with slotted lightweight concrete at one end and the dense concrete with foam inserts at the other end. Shu et al. (1979), Valore (1980) and Valore et al. (1988) found that the isothermal planes model gives a much closer approximation to the measured R-values (i.e., a large m-factor) than the parallel path (zone) model ($m = 0$).

For sheet metal constructions, differences between R_1 and R_2 were much larger than those for masonry constructions. The insight into heat flow patterns through sheet metal constructions led to the development of a method that uses a network of resistances related to the same representative area. Starting from textbooks on heat transfer the Swedish Research group (Johanesson and Aberg, 1981; Johanesson, 1981) developed a guide for design of highly insulated sheet metal constructions (Swedisol, 1981; Johanesson and Andersson, 1989). Some of these examples were recently verified with 3-D computer model (Bloomberg and Claesson, 1993). More recently, Trethowen (1995) proposed an alternative approach that incorporates contact resistance.

Recognizing these effects, the authors propose to use Equation (1) with the following m-factors:

$m = 1.0$ for wood frame constructions and insulating masonry blocks

$m = 1.4$ for ceramic masonry blocks and sheet steel construction, if the adjacent layer has thermal resistivity higher than wood (insulation)

$m = 1.8$ for concrete masonry blocks and sheet steel construction, if thermal resistivity of the adjacent layer is equal or lower than that of wood

To assess thermal transmittance when final contract documents are prepared, more precise methods may be required. For instance, in mixed and cooling climates, one may require the fifth level of accuracy in assessment of thermal transmittance, since this level involves transient performance under specific climate and use conditions.

The second component of thermal performance—air leakage—relates to the rate of air exchange across the building envelope. This component is directly proportional to air pressure differences across the envelope and to the area of air leakage openings.

The third aspect of thermal performance evaluation relates to the depression of temperature on the inner surface of a thermal bridge (TB). In cold climate, lower thermal resistance at these locations reduces the surface temperature.

The following aspects of thermal and moisture performance of thermal bridges in external walls require consideration:

- temperature depression on the surface caused by reduced thermal resistance
- additional depression of the surface temperature when the TB is located in the space with restricted air circulation (both cause an increase of the air film resistance)
- dust marking caused by the surface temperature depression (aesthetics)
- moisture condensation caused by the surface temperature depression (as it may increase the potential for mold growth and affect health of the occupants)

Typically, floor junctions in masonry construction and concrete decks connected to balconies experience problems with condensation because of significant depression of the surface temperature. When a reduction of convective air movement accompanies these temperature depressions (e.g., a closet or a staircase leading to a roof terrace), then mold growth can be expected.

3.1.3 CONTROL OF MOISTURE TRANSPORT—DESIGN WITH BARRIERS OR FLOW-THROUGH PRINCIPLE

When we talk about movement of water vapor, we think about moisture coming from indoors (in cold climates, or from the outdoors in hot and humid climates); but we rarely think about moisture from rain. Rain is typically excluded from the considerations, because protection against rain penetration is supposedly achieved by other elements of environmental control.

In one case study, rain penetration into the leaky masonry walls coincided with negative indoor pressure created by the HVAC (Lstiburek, 1995). The

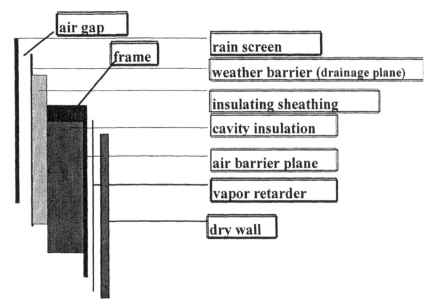

FIGURE 32. Schematic representation of environmental functions in a well-designed frame wall for cold climate.

rain was evaporated and vapor was carried inwards by the air movement. In another case, rain leaking through faulty window flashing accumulated between impermeable external cladding and vapor retarder causing damage to the inner part of the wall (Brown et al., 1997).

The environmental functions of the frame wall construction are schematically presented in Figure 32.

The first environmental function[56] shown in Figure 32 is a drainable air gap. It contributes to two aspects of moisture performance:

- control of rain penetration by reducing the intensity of rain carried across the air gap (ideally, no rain is carried across the gap when there is no pressure difference in the so-called pressure-equalized rain screen design)
- removal of residual rain and summer condensation of vapor (with wall surface temperature oscillating between day and night, the direction of thermal gradient in the outer layer of the wall varies and may cause moisture to move back and forth), unless it is removed through the drainable air gap

The drainage function of the air gap is often underestimated. Tradition introduced the air gap into frame walls and into masonry walls (double leaf,

---

[56]Each environmental function may or may not be fulfilled by one specific material or component of the building envelope system.

brick veneer). Only recently, when insulating these cavities with foams or fibrous insulation and causing moisture problems in renovated walls, we started to appreciate the importance of an air gap provided with the backup of the drainage plane.

Sometimes, in the rehabilitation of leaky, single leaf, masonry walls, one may consider creating a drainable air gap even behind the masonry wall.[57] Whether this air gap is ventilated or not is often much less important, since its main purpose is to function as the capillary break. Figure 33 illustrates the capillary breaking power of an air gap.

Figure 33 shows how a large difference in moisture transport was caused by a break of direct contact between two materials. The same moisture content was reached either in 20 or in 2000 hours. Obviously, this ratio is not a constant but depends on the capillary nature of both materials, yet it is clear that drainage is the most powerful method of moisture removal.

The next functional element shown in Figure 32 is the weather barrier, sometimes called the drainage plane to underline its function in moisture management. The weather barrier has two functions:

- draining rain water that may find its way to the inner surface of the air gap
- preventing ingress of air to avoid so-called wind-washing, when air enters and exits on the cold side, when the plane of airtightness of the air barrier system[58] is on the inside of the wall

One often forgets that the traditional weather barrier also functions as the drainage plane. In some walls, particularly when the external insulation finishing systems (EIFS) provided sufficient resistance to air ingress, the need for the drainage plane was only rediscovered, as the second line of defense, when faulty window connections introduced water into the building envelope. The need for the second line of defense was recently highlighted in a series of papers (e.g., Brown et al., 1997).

The next functional element (Figure 32) is a thermal insulating sheathing placed on the cold side of the wall frame. Its role is particularly important in steel frame walls or highly insulated frame walls. In this case, the frame can significantly reduce the efficiency of the thermal insulation.

The next functional element, following the cavity insulation is an air barrier plane. Section 5.4 of the 1995 National Building Code of Canada describes performance requirements for an air barrier system. The plane of airtightness must reduce the air leakage to less than $0.02 \, l/(s \, m^2)$ when tested at a 75 Pa pressure differential.

Finally, the last functional element, the vapor retarder, has received a dispro-

---

[57]Michel Perault, presentation at Building Envelope Seminar in Ottawa, 1996.
[58]The air barrier systems are discussed in Chapter 10.

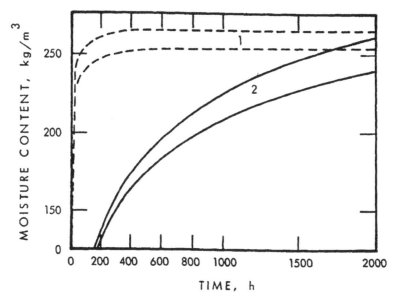

FIGURE 33. Moisture exchange between a dry aerated concrete specimen and a wet gypsum specimen, at two levels of moisture content in gypsum and two different conditions at the surface: (1) Samples in contact, (2) samples separated by 1 mm air gap (Bomberg, 1974).

portionate amount of attention in construction practice. One explanation is that the vapor diffusion is one of the few moisture transport mechanisms that are easy to calculate. These calculations can only be used to establish the onset of condensation. As the condensation process makes them invalid, they cannot be used to determine the quantity of condensed moisture.

Often, to be on the secure side, practitioners do not allow any condensation within the building structure. We denote this type of approach as moisture design with barriers. Another approach, called a flow-through principle, requires performing experiments or calculations to estimate how much moisture will accumulate within the building envelope during the worst season. Then, one must assess the potential effects of this moisture on the structure durability. If the durability is not adversely affected, one must check if the drainage and drying capability over the remaining part of the year will exceed the condensed amount. Moisture is allowed to accumulate during the heating (or cooling) season, but it is not allowed to accumulate from one year to another.

Figure 34 illustrates these two different design principles with an analogy to a barrel. No liquid is delivered to the barrel and none is flowing out (design with barriers). The same quantity flows into the barrel as it flows out and the balance does not change (flow-through on the yearly basis).

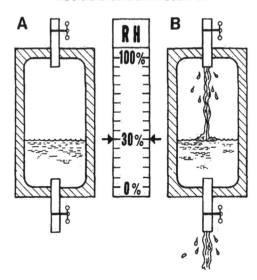

FIGURE 34. Two approaches to moisture design: (1) design with barriers, (2) design with flow-through principle.

Note that the moisture flow is seldom one-dimensional, and the flow-through does not mean the vapor diffusion through the building component. Actually, the more drainage occurs during phase changes of moisture[59] entrapped within the enclosure, the more efficient is the flow-through design. The name flow-through was selected to emphasize an important aspect of the design, called moisture tolerance or "forgiveness" of the building system. A well-designed building system "forgives" us, when, because of mistakes or unforeseen circumstances, some moisture enters the structure. There is no damage to materials and the system allows for subsequent drainage and drying.

3.1.4 MOISTURE EFFECTS—MATERIAL DURABILITY

The building envelope must perform, retaining its structural integrity, while separating the interior and exterior environments. Of all environmental conditions, moisture poses the biggest threat to integrity and durability of materials in building envelopes. Many construction materials contain moisture, most notably, masonry or concrete. These materials demonstrate excellent performance as long as the moisture does not compromise their structural or physical integrity. However, excessive moisture jeopardizes both the material and its functionality.

---

[59]Phase changes may occur on daily basis because of solar or night radiation.

Consider, for example, the ability of a material to withstand, without deterioration, natural periods of freezing and thawing. As already mentioned, the frost durability is not a material characteristic but a complex property which depends on the material, the construction system and the environment. For instance, in one school building, only the outer surface of external clay-brick protrusions showed freeze-thaw spalling. These protrusions were more exposed to driving rains and the surface temperature of the bricks was slightly lower, compared to the plain facade where no spalling occurred. Both of these conditions contribute to increased risk for freeze-thaw damage.

One may also observe interaction of temperature and moisture in other types of moisture originated damage e.g., corrosion and mold growth. Rate of corrosion of metals exposed to air varies with both the surface temperature and relative humidity of air (Grodin, 1993). Mold growth requires coincidence of both certain temperature and humidity ranges [temperatures above 5°C and relative humidity above 80% (Hens, 1992)].

In effect, the design for durability may require many technical and cost considerations related to the environmental control of the envelope, buildability, defects arising during construction, inspection (commissioning), maintenance, repair or replacement, and calculation of the life-cycle cost of the structure.

This discussion would not be complete without addressing a simultaneous effect of thermal bridges and workmanship on thermal performance of fibrous insulations.

3.1.5 EFFECT OF TB AND WORKMANSHIP ON THERMAL PERFORMANCE OF FIBROUS INSULATIONS

How much air movements can affect thermal performance of fibrous insulation depends on air pressures within the building element and its surroundings, air permeability of the insulation, and the airtightness of joints between materials and building elements. Bankvall (1986, 1986a) showed a dramatic reduction of thermal performance of mineral fibre batt exposed to air movement along the insulation surface and a paramount effect of workmanship on thermal performance of the wall.

As Bankvall dealt with a hypothetical case of poor design and poor workmanship, Brown at al. (1993) studied a specific case of workmanship faults. Each vertical corner of the wall cavity was partly unfilled (3 and 6 percent of the area were selected for testing). The wind protection was applied on both sides of the mineral fiber insulation and no continuous and interconnected air spaces and gaps were simulated. This kind of test condition appeared to come closer to workmanship faults found in actual walls. Figure 35, quoted from this research, showed that with a large difference in temperature, the reduction of wall thermal performance may be as high as 50 percent.

Silberstein and Hens (1995) analyzed the significance of proper design of the ventilated air spaces. It is important to underline that similar to a modern wall construction, the roofing deck must also be constructed as the airtight structure. It was shown that for an airtight roof deck structure, and air velocities observed in ventilated cavities, the effect of air ingress into the insulation is insignificant.

There is no contradiction between the results of these two studies. Brown et al. (1993) stated: "Since the measured thermal resistance of walls with 0% defects agreed with predicted values, it is evident that the material is performing as expected; consequently, the issue of installation practice needs to be examined." The authors also stated: "It appears that the convective flow was initiated in the cross-section between the hot/cold pair of air gaps and then spread throughout the rest of insulation. A contributing factor is that the air permeability along the MFI product, the manufacturing plane, is much higher than across the product."

This explanation follows findings of Wilkes et al. (1991), who showed the significance of convection initiators on the onset of convection. This closes a loop of practice and understanding. The transition zone to fully developed natural convection in horizontal layers was already implied by Wilkes and Rucker (1983), but with the more recent research of Wilkes (1991) and Brown et al. (1993) the role of convection initiators was first understood.

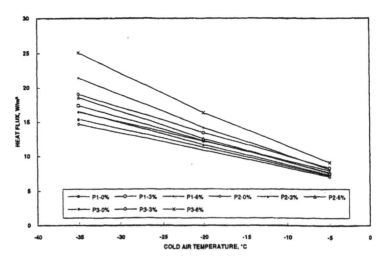

FIGURE 35. Reduction in thermal resistance of the frame wall insulated with MFI batts caused by workmanship faults, namely corner spaces along the studs are not filled to the extent of 3 percent or 6 percent of the cavity cross section. (From Brown et al., 1993.)

In practice, reduction in thermal performance of insulation caused by convection was observed at NRC by Woolf et al. (1966) on wood-frame walls and Sasaki (1971) on steel-stud walls. The latter paper discussed the effect of air gaps caused by a 6 mm-lip on the flange of the steel stud. Brown (1986), testing 14 different configurations of sheet steel walls, indicated similar reasons for poor performance of some of the tested sheet steel walls. So it is not the occurrence of thermal performance reduction, but the understanding of its mechanisms that constitutes the progress of the last thirty years.

3.2 Indoor Environment: Comfort and Air Quality

Ventilation is required for the health and comfort of the occupants. Ventilation is the process of supplying fresh air and removing "old" air to maintain adequate indoor air quality and to supply air for combustion devices.

Natural ventilation through random and discrete openings such as operable windows, doors, ductwork, or holes is not adequate because of the lack of consistency of the driving forces. The infiltration and exfiltration rates vary substantially due to the influences of wind pressures, stack pressures, and pressures induced by air-consuming devices. Some areas of a house can have adequate air change at one moment and inadequate at the next. It is important to recognize that relying on random leakage openings and the effects of wind and stack effects to provide the required air change does not ensure dilution when most needed. In small buildings, the variations in building airtightness are enormous. Leaky buildings are often 2 to 3 times leakier than tight buildings. Compounding these concerns is the difficulty in predicting how tight a building will be when built using conventional construction practices. In effect, for health reasons, mechanical ventilation is a requirement in all buildings, tall or small.

Mechanical ventilation involves the provision of a controlled driving force to remove and supply ventilation air through either deliberate, discrete openings or through random openings. This driving force can be provided on a continuous basis, or on demand, to remove specified pollutants.

Sources of pollutants and odors can be controlled at the point of generation (source control); for instance, the direct venting of combustion appliances and installation of range hoods in kitchens. These sources can also be controlled by prohibition (exclusion) by the building officials enacting appropriate regulations (for example, banning non-vented, kerosene space heaters), and requiring the pressurization of crawl spaces or basements to exclude radon gas.

Pollutant concentrations are one of many components affecting the indoor environment of a building. The indoor environment also involves comfort factors such as temperature, relative humidity, air velocity, physical stress

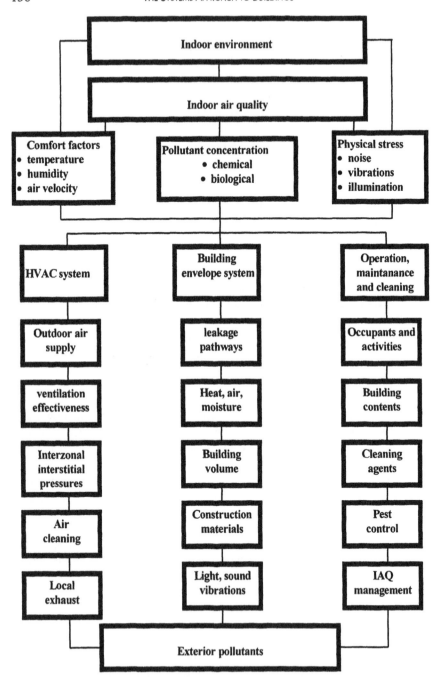

FIGURE 36. Factors affecting indoor environment.

(noise and lighting) and psycho-social factors (personal relationships and work stress) and chemical, particulate, and biological concentrations. The complexity of the interrelationships between the indoor environment, indoor air quality, and comfort factors is illustrated in Figure 36. The relationships of factors shown in this figure involve health, safety, durability, comfort, and affordability concerns as well as questions about construction performance (warranty).

3.3 HVAC Considerations

The design of the building envelope provides a basis for design of the heating, ventilating, and air conditioning (HVAC) systems. At the same time, the design, installation, and operation of the HVAC system affects all aspects of indoor climate and building envelope durability. Air movements induced by HVAC may affect pollutant migration, rain penetration, condensation, and drying of moisture within building cavities, i.e., durability of building components.

As long as buildings were leaky and poorly insulated, the effect of HVAC systems on air pressure was insignificant. There was no need to understand air movements in the building, more than providing a necessary supply of fresh air. This is not the situation today. Now, we have well-insulated, airtight buildings and an increased incidence of health problems (mold/microbial contamination), and deficient long-term performance (metal corrosion and other moisture originated deterioration). Airflow carries moisture which affects materials' long-term performance (serviceability) and structural integrity (durability), airflow impacts the spread of smoke in a fire situation, the distribution of pollutants, and the location of microbial reservoirs (indoor air quality). Air exchange affects energy for space heating or cooling.

The key to these real or potential problems is understanding of air pressure fields in the indoor and interstitial spaces of the building envelopes. Today, the understanding of air movements in the building is a necessity. The determination of air pressure differences, however small and difficult to measure, is needed to establish performance of the building as a system.

3.3.1 STRATEGY TO CONTROL AIR PRESSURE

A strategy to control air pressure in the building space includes the following steps:

1. Enclose the air space
2. Use controlled mechanical ventilation

3. Control air pressure fluctuations induced by HVAC system operational conditions
4. Control air pressure gradients induced by HVAC system operational conditions
5. Eliminate interconnected internal cavities communicating with HVAC systems
6. Review the building mezzoclimate for differences in wind and solar shading conditions

To control the air pressure field, you must first enclose the air space. The next step is to quantify the degree of airtightness for any building envelope. The National Building Code of Canada (Swartz, 1995) made the requirement of an air barrier system mandatory and provided performance objectives for testing and evaluation of these systems.

The air barrier controls flow of air through external envelopes of the building. However, the effect of pathways created by external cavities and interconnected internal cavities communicating with HVAC systems on performance of building systems is seldom recognized. The significance of these elements, mostly neglected in the traditional analysis of air pressure fields, will be illustrated in the few examples selected from case studies (Lstiburek, 1992, 1994, 1995).

The following two examples involve the effect of pathways created by external cavities and interconnected internal cavities communicating with HVAC systems. The first one (Figure 37) involves a leaky return duct (actually it was the housing of an air handler) enclosed within an interstitial space. The second one (Figure 38) involves a plenum return ceiling communicating with an exterior wall.

Figure 37 illustrates a demising wall communicating with a leaky return duct in a building located in a hot, humid climate. The leaky return duct created a negative pressure. Since the cavity in the demising wall is connected with the furring space in the exterior wall, the interconnected cavities can extend the effect of the leaky duct for a great distance. In the actual case study, moist outside air was drawn into the building cavities in spite of the positive pressure in the interior space.

Figure 38 illustrates a plenum return ceiling which is not sealed at the exterior perimeter wall in a building located in a cold climate. Plenum return ceilings operate at negative pressures, which may range from 1 to 2 Pascal (negative to the interior space) to 20 to 30 Pascal (negative to the interior space). When the plenum return ceiling is also negative to the exterior, outdoor air can be drawn into the plenum return through the exterior wall assembly. The error in the design shown in Figure 38 had caused additional problems, as in the studied case, the exterior wall cavity was also connected to the crawl space. Moisture and pollutants were drawn into the return air plenum.

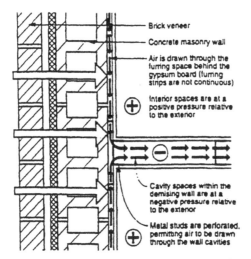

FIGURE 37. Negative pressure created by the leaky return duct draws moist external air into the wall cavity of the building located in a hot and humid climate.

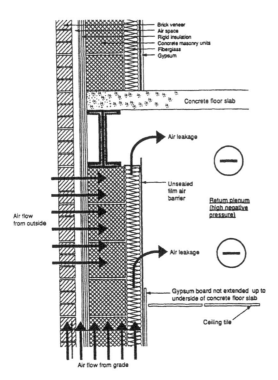

FIGURE 38. The exterior wall cavity was connected to a foundation assembly. The error in design (unsealed gypsum) permits drawing air from below grade to the return plenum.

3.3.2 BUILDING ENVELOPE AND HVAC
SUBSYSTEMS INTERACTION

The following two examples illustrate the significance of ducted distribution systems on interior air pressures. The first one (Figure 39) involves the leakage of supply ducts installed in an exterior space (vented attic). The second one (Figure 40) involves a facility with inadequate provisions for return air.

Figure 39 illustrates a facility located in a hot, humid climate with leaky supply ducts located outside of the conditioned space in a vented attic. Air leaking out of the supply ducts depressurizes the conditioned space, inducing the infiltration of exterior hot, humid air.

Figure 40 illustrates a facility located in a cold climate with inadequate provision for return air. When interior doors are closed, individual rooms/spaces can become pressurized with respect to common areas. The common areas, in turn, become depressurized. If atmospherically vented combustion appliances (such as fireplaces and gas water heaters) are located in the common areas, the negative pressure in these regions can lead to spillage and backdrafting of combustion appliances. In the pressurized rooms/spaces, the forced exfiltration of interior (typically moisture laden) air can lead to condensation and moisture-induced deterioration problems.

In most mid-rise and high-rise buildings, the stack effect airflows typically dominate the HVAC system airflows. Stack effects are shown schematically in Figure 41. Note that the majority of the air pressure drop is taken by the exterior building envelope at the top and bottom of the building. Airflows from the lower units and floors, up the elevator shafts, stairwells, and service pene-

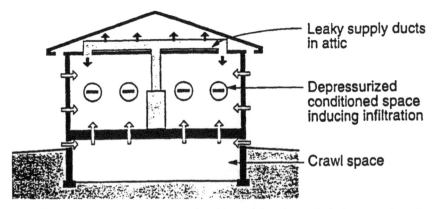

FIGURE 39. Negative pressure created in the conditioned space by leaky supply ducts located in the attic of the building located in a hot and humid climate.

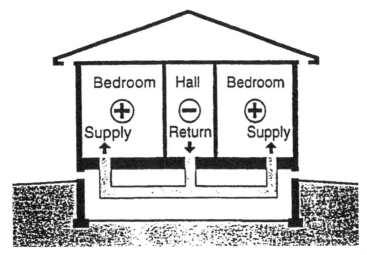

FIGURE 40. Pressure differences in the indoor space created by inadequate provision for return air when doors are closed.

trations to the upper units and floors. These stack-effect-induced airflows are often responsible for pollutant migration, odor problems, smoke and fire spread, elevator door closure problems, and high thermal operating costs.

Better conditions are achieved by sealing units from corridors and by isolating corridors from elevator shafts (vestibules); airflows caused by stack ef-

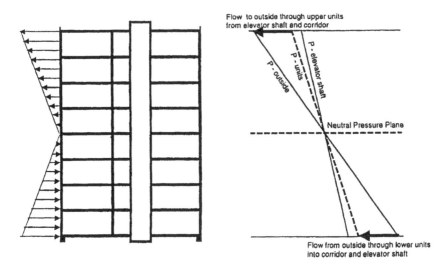

FIGURE 41. Air pressure distribution in a mid-rise building.

fects are significantly reduced. The forces acting on this eight-story building have been reduced by "compartmentalization." In essence, this building behaves in a similar fashion to eight, one-story buildings located on top of each other. The pressure drops are now taken across the corridors and elevator vestibules, not the exterior building envelope. This results in a safer building with respect to smoke and fire control. IAQ problems are reduced and energy efficiency is greatly enhanced.

Figure 42 illustrates a controlled ventilation system that is recommended for the eight-story tower. The main features are:

- Units are ventilated individually through exterior walls.
- Units are compartmentalized and isolated from corridors and shafts.
- Corridors and stairwells are pressurized via a smoke control system. The existing duct system for corridor air can be utilized.

In multistory office buildings, compartmentalization was traditionally accomplished by placing elevator banks in vestibules, which isolated them from the remainder of a floor. The vestibules acted similarly to air locks when the doors to the floor were closed, preventing air from a floor rushing up the elevator shafts. The utilization of vestibules around elevator banks creates a circular or "donut" zone of pressure control (Figure 42) at each floor.

Duct leakage can inadvertently cause communication between seemingly isolated spaces. Figure 43 illustrates a storage space containing printed mate-

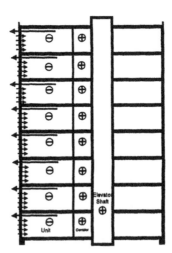

FIGURE 42. Air pressure distribution in a mid-rise building where units are compartmentalized, and isolated from corridors and shafts and ventilated individually through exterior walls. Corridors and stairwells are pressurized via a smoke control system and via a fire alarm control.

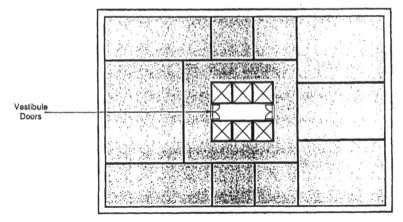

Vestibule
Doors

FIGURE 43. Shaded area is pressurized relative elevator vestibule and exterior wall interstitial cavities.

rials and an operating print shop. The storage space is maintained under a negative air pressure with respect to the rest of the facility by the operation of an exhaust fan. However, a return air duct passing through the storage space was found to be leaky resulting in high levels of volatile organic compounds (VOC) being drawn into the air handling system serving a neighboring office space. The result was transmission of VOC from the storage space to the office space via the HVAC system return duct leakage and health complaints in the office space.

Hallways and corridors can cause an extension of pressure fields throughout a building. A typical hotel room ventilation system may have a bathroom exhaust operating on a continuous basis via a rooftop mounted exhaust fan (which also serves for other bathrooms). Make-up air for this bathroom exhaust is typically provided through the exterior wall via a unit ventilator or packaged terminal heat pump (PTHP). In the studied case, the design assumed that 60 cfm out through the bathroom is offset by 60 cfm in through the unit ventilator or PTHP. Although the unit ventilator or PTHP does not run continuously, an intermittent imbalance of 60 cfm is not considered a problem.

Consider thirty hotel rooms on a floor served by a single corridor. This corridor becomes a large duct connecting all rooms on the floor. With 30 exhaust flows of 60 cfm each, if they operate continuously an 1800 cfm exhaust is created on the floor. Unit ventilators or PTHPs are only operating on a 20 percent duty cycle (i.e., 80 percent of the units per floor are not operating at a given time). The supply air is only 360 cfm (6 operating unit ventilators of PTHPs at 60 cfm each). The flow imbalance per floor is 1440 cfm, which is sufficient to depressurize the entire floor.

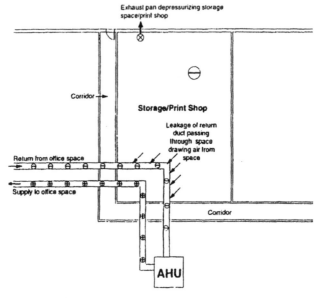

FIGURE 44. Duct leakage can cause communication between seemingly isolated spaces. In the case study, leakage from a return duct passing through the storage space and the print shop introduced a pollution transfer despite an exhaust fan depressurizing this local area.

In hotel facilities located in hot and humid climates, the negative pressure field created in this manner is the single, most significant reason for mold, odor, and moisture damage. This observation highlights that air leakage/pressure relationship is the key to the interaction of the building envelope and the HVAC system. To design and build safe, healthy, durable, comfortable, and economical buildings, we must control the pressure fields.

4. CONCLUSION: THE NEED FOR A HOLISTIC APPROACH TO DESIGN

To design long-term performing and durable building envelopes, the micro-environment within the building envelope must be controlled. To this end, one needs to deal with heat, air, and moisture transport collectively. As shown in the discussed examples, we need to mix the experience and judgment based on an understanding of how the building envelope functions with the analytical evaluation of selected performance factors (based on testing, computer models, and/or simplified calculations).

The above discussion showed that heat, air, and moisture transports are inseparable phenomena. In effect, to address interactions and trade-off between control of heat, air and moisture transports, the building system must be analyzed as a whole at the same time each of its components is analyzed. We often simplify the design process by examining these issues separately and ascribing control of each phenomenon to a particular material; yet, somewhere in the design process, we must examine their interactions. Particularly significant are the interactions between building envelopes and mechanical systems. They impact health and safety (indoor air quality, smoke and fire spread), durability (moisture transport and accumulation), comfort (temperature, relative humidity, odors) and operation/maintenance (energy costs, minor and major repairs, housekeeping). This review highlighted limitations of the traditional approach and difficulties in developing accelerated testing of durability.

Current building codes stress the traditionally acceptable solutions while product evaluation based on tradition restricts introduction of new construction products. Architectural design process fails to examine the relationships between input coming from many disciplines: architecture, structural and mechanical engineering, fire protection, acoustics, and interior design. One needs to develop models of holistic approach to design and quality assurance of the environmental control function. One such model is proposed in another chapter.

CHAPTER 9

Design and Performance of Roofing Systems with SPF

O VER THE YEARS, many academic and research organizations attempted to develop a comprehensive method for evaluating the performance of all components and systems in a building. Though the development of a universal performance-evaluation system so far has eluded professionals, some limited and focused attempts were more successful. One of them was started by the Roof Systems Research (RSR) Committee of the polyurethane contractor's organization, which preceded the current SPF division of the SPI Inc.

1. PERFORMANCE OBJECTIVES FOR SPF ROOFS

The RSR committee, whose members were mostly nonindustry volunteers, recognized that the SPF roofs must be analyzed as a system. While this was done for fire and wind uplift evaluations, it was not the case for assessment of other aspects of system performance. Therefore, the committee defined its purpose as follows:

- to recommend test methods and performance criteria for SPF and liquid applied weatherproofing coating used in roofing
- to insure that these test methods and criteria are applicable to all roof systems with consideration of intended service
- to identify factors affecting SPF roof performance which must be considered during design, installation, and in-service environments

An extract from RSR committee work is presented in Table 41 using current terminology (see Appendix A4). The symbol " ✕ " indicates that the RSR committee identified the test method with or without acceptance criteria. The symbol " ♦ " indicates that the RSR committee postulated the need to develop a test, but the test and/or acceptance criterion was only recently developed. Finally, the unresolved issues are also identified in Table 41.

Some performance objectives that were proposed by the RSR committee for specific SPF applications are not discussed here.

167

Table 41. Performance objectives for SPF roofs included
in the industry-wide QA system.

Performance Objective	Test Method	Test Criteria
Adhesion and static wind uplift strength of the roofing system	♦	♦
Dimensional stability of the SPF	♦	missing
Fire resistance (considerations for materials and systems)	×	×
Abrasion, impact and puncture resistance	×	×
Water vapor permeance (transmission through the roofing system)	×	♦
Differential structural movements, thermal movements, and thermal shock resistance	×	♦
Long-term thermal resistance	♦	♦
Moisture accumulation caused by presence of thermal gradients	♦	♦
Dynamic wind uplift	(?)	missing

1.1 Adhesion and Wind Uplift Strength of the Roofing System

Adhesion/cohesion strength is amongst the main considerations in the design of any roofing system. Traditionally, only static wind-uplift tests were conducted on the mock-up assembly. Hurricanes like Alicia in Houston, Texas in 1983, or Hugo in South Carolina in 1990 showed that roof assemblies rated FM I-90 (rated for 4.3 kPa or 90 psf), which corresponds to 150-mph wind, failed at wind speed 40 to 60 mph.

Today, we realize that there is a need for both static and dynamic evaluation of wind uplift. For instance when discussing the application of limit states design for durability of buildings, Allen and Bomberg (1997) analyzed metal roofing subjected to fluctuating high wind pressures, where fatigue failure occurs in the sheet metal surrounding the connectors. Because of the complexity of structural behavior, the standard fatigue curves based on component stress are not applicable. Testing is required for each design for both fatigue and ultimate static resistance. Baskaran and Dutt (1995) showed results of such tests for particular design, where the fatigue resistance is expressed as a percentage of the static resistance. In this example, the fatigue resistance for 1000 cycles of stress range is approximately 20 percent of the ultimate static resistance. This indicates that low-cycle fatigue is a likely failure mechanism for metal roofing.

How does SPF look in this respect? Normally, the adhesion of SPF to clean steel or concrete surfaces (see Chapter 2) exceeds the static wind-uplift forces by a factor of 100. Typically, a continuous layer of the SPF provides sufficient airtightness and reduces the differential wind loads that could cause a local tear of discontinuous membranes. However, adhesion/cohesion problems may occur when an existing built-up roof system is recovered with the SPF. Since the poor adhesion within the existing BUR may not be detectable by visual inspection, either test cuts or test cuts combined with the field test of adhesion/cohesion (see Section 5.2) are recommended.

1.2 Water Vapor Permeance and Moisture Accumulation

Often, one confuses the need to evaluate characteristics of the whole roof system with that of the SPF product. While the water vapor permeance of SPF is an important property, it has only a minor significance in the assessment of moisture performance of the whole SPF roof. Moisture accumulation in the roof depends on its design, indoor and outdoor conditions, permeance of elastomeric coating, permeability of internal finishes, and penetrations in the roof membranes. The water vapor permeance alone does not provide enough information to assess the moisture performance of the roofing system.

As previously stated, the roofing materials may contain moisture and function very well. Only when the moisture accumulation affects the structural integrity or performance of the materials and components, does the presence of moisture become a concern. In this respect, Kashiwagi and Moor (1986), realizing that most vapor drive is from the inside out, highlighted the importance of "breathing through coatings," i.e., vapor permeable coatings. As a criterion for a breathable coating, we will require that its water vapor permeance be higher than that of the 25 mm- (1 in) thick SPF layer.

The transmission through and the accumulation of moisture in SPF may be calculated with a verified computer model. The material characteristics for the actual product are, however, seldom known and similar material data (from a database) are used. It is therefore advisable to check the material characteristics by performing a benchmark test and comparing measured and calculated results. It is recommended that a test on moisture accumulation caused by a thermal gradient (Chapter 3) be used as the benchmark test. When the computer prediction has been verified against moisture accumulation in the benchmark test, the computer model can be used for selection of a coating layer for any specified set of environmental conditions.

1.3 Long-Term Thermal Resistance

Chapter 3 describes and applies new methodology to establish long-term thermal resistance (LTTR) of the SPF with permeable coatings. Using the LTTP for a 15-year period of energy calculations, SPF contractors may provide the bidding documents that include precise prediction of future energy use for roofing systems or walls filled with SPF.

Chapter 10 (Section 4.4) documents the agreement between predictions obtained with the discussed methodology and the full-scale measurement performed on walls exposed under field conditions. For a frame wall filled with 76-mm (3″) thick MD-type SPF, the predicted LTTP was 3.3 percent higher than the value measured after five years of exposure.

The same section analyzed the effects of different elastomeric coatings on long-term thermal performance. It found, in two cases, that multilayered, elastomeric coatings could act as impermeable barriers causing significant retardation of the aging process. These two cases were: Hypalon covered with modified urethane and built to thickness of 1.2 mm (50 mil) and Butyl basecoat with Hypalon topcoat and catalyzed urethane [built up to 2 mm (80 mil) thickness]. In both cases, the LTTR of the SPF was equal to that of PIR faced with an impermeable foil.

1.4 Other Considerations: Abrasion, Impact, and Puncture Resistance

Resistance to abrasion from traffic and wind-borne particulate needs also to be considered during the roof design. Normally, the requirements for resistance against abrasion, impact and puncture, wear and tear from foot traffic etc. are much higher on the areas designated for traffic and surrounding the HVAC equipment than on the remainder of the field. Sometimes, these areas are given special consideration and marked with a different color on the roof surface.

Resistance to abrasion, impact and puncture, were adequately addressed in Chapter 4 in which the standard tests for elastomeric coatings are discussed.

1.5 Unresolved Performance Objectives

Some performance objectives that were identified by the SRS committee ten years ago are still on the priority list now, as much as they were then. The two most important considerations are:

- dimensional stability under thermal gradient
- differential structural and thermal movements, thermal shock resistance

1.5.1 DIMENSIONAL STABILITY OF THE SPF

The SPF layer must exhibit sufficient dimensional stability so as not to cause separation or pass delamination during the expected service life. As discussed in Chapter 2, Section 4.1, however, the existing test does not provide enough information about dimensional stability of SPF under field conditions. Development of a dimensional stability test, which employs thermal gradient remains the priority issue.

1.5.2 DIFFERENTIAL STRUCTURAL AND THERMAL MOVEMENTS

The SPF roof shall withstand structural movements and sudden or frequent changes in temperature without surface cracking. This issue has not been addressed by the chemical industry, that examines material performance in laboratory conditions where specimens are alienated from all the system interactions that affect their field performance. The RSR committee stated: "The amount of movement experienced by spray-in place polyurethane foam roof systems in service has not been measured. Such measurements are needed or should be estimated using appropriate methods." The situation is the same ten years later.

1.6 Engineering Solution for SPF Movement Bridging

One of two issues identified ten years ago by the RSR committee but still unresolved today is the quantification of movements experienced by the SPF roofing system, and this lack of knowledge prevents us from developing an appropriate test method.

Therefore, to build durable roofs, we need an engineering solution where materials with ability to withstand movements while retaining their continuity as the barrier are used to aid the SPF layer. This monograph recommends the use of elastic, waterproofing, membrane strips located under the SPF layer, over joints, junctions, and other locations where significant structural movements are expected. This monograph uses a new term for a membrane that controls Water, Air, and Movements, namely a (WAM) membrane. Performance objectives and material selection for WAM membranes are discussed in the appendices.

2. GENERAL CONSIDERATIONS IN SPF ROOFS

A database involving more than 1,000 SPF roofs located in six geographical areas was created at the Arizona State University. Since 1982 this database has continuously been updated. Analyzing defects observed in these roofs (pass delamination, blistering, moisture in the foam, ice cracking, leaking at joints and penetrations) and asking several questions to the owners and facility managers, Kashiwagi and Moor (1986) defined a "successful" SPF roof (Figure 45).

Kashiwagi and Moor (1986) required that a SPF roof be built with HD-type, minimum 50 mm- (2 in) thick layer of SPF and designed for 20 years of service life. The design should be based on a "complete understanding of all factors affecting the performance of the system in a defined environment." If the knowledge of the system operation in its environment is limited, one "de-

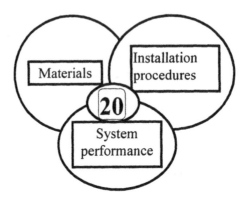

FIGURE 45. "Successful" SPF roofing system (Kashiwagi and Moor, 1986).

fines the factors affecting the system as completely as possible," and supports it with experimental results or with data obtained on existing "successful" polyurethane roof systems.

Whenever a delamination between passes occurred, two application-related factors were observed:

- The surface of the layer beneath was usually smooth, and slightly burned from UV rays.
- The thickness of the upper pass of SPF was inconsistent.

These observations led to changes in the application technique. The application practice was changed to build full thickness of the SPF roof in one day with a few passes, each between 1/2 and 1-1/2 inches thick.

Recommended thickness of the SPF pass varies. In a cold environment, passes thinner than 1/2 inch may lose too much heat to the environment, and insufficient vaporization of the blowing agent may upset cell growth. Passes that are too thick may result in excessive heat buildup that reduces retention of the blowing agent in the cell-gas resulting in poorer long-term thermal performance. Typically, for manual application, the maximum pass thickness is 25 mm (1 in) and is 40 mm (1-1/2 in) for robotics applications. Since the maximum pass thickness depends on the foam reactivity profile (e.g. cream time), foam density (mechanical properties of the fresh foam), as well as the installation technique (manual, robotics), the manufacturer must be consulted for specific recommendations.

HD-type foams are usually applied with passes thicker than the MD-type foams because in vertical applications, the lateral creep (movement that occurs during the foam growth) may have a bigger impact on inter-pass adhesion.

Kashiwagi and Moor (1986) recognized that moisture performance is one of the critical factors in the SPF roof design. Actually, it is not moisture presence itself, but the effect of moisture on the balance of forces and movements on inter-pass surfaces. These aspects of SPF performance are discussed in Chapter 2, Section 6.4.

Typically, general considerations in design of roofing systems include:

1. Selection criteria
2. Required properties of the roofing foam
3. Required properties of the protective coatings
4. Substrates, primers, and the surface preparation
5. Drainage and penetrations
6. Roof-mounted equipment and walkways
7. Maintenance
8. Expansion joints and roof details

Many of the above considerations are addressed in the material specification for SPF (Appendix A4) and checklists (Appendix A7). Other considerations are discussed below.

2.1 Substrates, Primers, and the Surface Preparation

SPF can be applied to various substrates including concrete, wood, metal and some existing roof membranes providing that surface is clean, dry and free of oil and grease, unaged metal or uncured solvent-based materials. Primers are sometimes recommended to facilitate adhesion, particularly for surfaces of new concrete and wood. Typically, chlorinated rubber and urethane can be used for most substrates; epoxy for concrete, wood, metals and gypsum; neoprene for asphalt base sheet, concrete, wood, ferrous metals, and gypsum.

All surfaces to which SPF is applied must be clean of oil, grease, and form release agents and must be sufficiently dry (at least 28-days-old concrete). Moisture contained by concrete or wood may react with the fresh SPF producing carbon dioxide and reducing the bond to the substrate. All joints holes and openings that exceed 6 mm (1/4 in) should be grouted or caulked with elastomeric sealant and/or taped. Sometimes, to enhance adhesion, a primer may be recommended (see NRCA, 1996).

For existing metal substrates, the cleaning usually requires removing of grease and oil with a solution of tri-sodium phosphate and rinsing with fresh water. Areas of rust should be primed with rust inhibiting primer (e.g., zinc chromate, see NRCA, 1996). Soft mastic, paraffin-containing cold-process adhesives and other repair materials, which are known for poor adhesion to SPF, are also to be removed. Then the exposed surfaces are to be washed under a pressure of 13,800 kPa (2,000 psi).

2.2 Drainage and Penetrations

All SPF roofs are designed and built with a positive drainage. The slope of 6 mm (1/4 inch) per 305 mm (1′) is a minimum expected to satisfy the requirement that no standing water is present on the roof 48 hours after the rain. To assist drainage of the existing roofs, additional drains should be considered for areas with ponding water. To promote drainage, SPF can also be applied to built-up low areas, creating taper or built-up with rigid insulation boards.

The SPF on penetrations, equipment curbs and parapet walls should provide a smooth transition between the horizontal and vertical surfaces. The foam should be sprayed at least 150 mm (6 in) above the roof surface and the protective coating should extend at least 100 mm (4 in). A separate metal counter-flashing, protective covering, or cladding is required above this height, and above the parapet wall surface.

Metal roof jacks should be installed to isolate projections from the roof system when excessive heat or vibration is expected.

Where movement is expected in the roof construction, the expansion joints, flashing collars, and counter-flashing should be installed to accommodate these movements.

Electrical wire, cables and refrigerant lines should be extended through the roof in conduit. After the conduit is securely fastened to the roof deck, the SPF can be used to flash the conduit to the roof system.

Wall flashing and penetration flashings are normally sprayed in conjunction with the SPF application. Pitch pans, equipment support and curb-basis, are normally encapsulated in the SPF and coated, eliminating the need for metal flashings. SPF is tapered into the basin of the drain and coated. Examples of details with flashing, counter-flashing, pipe, and vent penetration are provided by SPFD guides. Nevertheless, the contractor's warranty should specify standard details that are recommended by the foam manufacturer.

2.3 Roof-Mounted Equipment and Walkways

Use of SPF may eliminate most complicated flashings required to seal around the equipment. Nevertheless, one must pay attention to the following considerations:

- Does the structural design allow for concentrated loads or will these loads cause deflections leading to a possible ponding of water?
- Roof-mounted equipment should not be located in valleys and low areas as it may reduce roof drainage.
- Spray-applied crickets should be added to the uphill side of large units to eliminate ponding.

- Pipe water discharged from HVAC equipment should not lead to roof drains.
- Space around pipes, curbs, ducts, and equipment should be provided for future maintenance and inspection of mechanical systems.

For construction details that can be used for placing heavy equipment, consult NRCA (1996).

Usually, walkways are not required for occasional traffic. If heavy traffic is anticipated, walkways can be installed. Typically, walkways connect with the shortest possible route such areas as 0.9 m- (3 ft) wide area around the perimeter, roof-mounted equipment, roof hatches, access ladders, and doors leading to the roof. The following types of walkways are typically used:

- additional application of one or more coats of protective coating (different color), used alone or in combination with additional ceramic granules and/or reinforcing fabric (the latter is placed between two coating layers)
- additional coating with embedded fabric or granules
- ½-inch-thick, 30-inch-wide, "breathable" walkway pad, spot-adhered with coating, caulking, or adhesives, interrupted at each 6 ft with a 2 inch gap to allow surface water drainage
- made of SHD-type SPF (compressive strength higher than 380 kPa, 55 psi)

Note that if granules are used, they are embedded in the additional coat (not in the standard coating); if the reinforcing fabric is used, it is usual to place it between two additional coats.

2.4 Maintenance

It is recommended that roof inspections be performed in spring and fall of each year. After cleaning, fill the punctured area with an elastomeric sealant or caulk of the same generic type as the coating. Tool the colored sealant to extend at least 1 inch beyond the damaged area.

Extensive repairs should be made by a qualified contractor. If the roof is under warranty, the company must be notified of any additions and alterations.

3. ROOF DECK TYPES

Several types of roof decks listed in the NRCA Roofing and Waterproofing Manual may be divided into two classes:

- SPF is applied directly to the existing deck.
- SPF is applied on the overlay board (adhered or mechanically fastened).

The following roof decks may be used for direct application of the SPF:

- Structural concrete (cast in place or precast), if concrete is fully cured and has passed a dryness test.
- Wood-plank and wood-panel decks, if protected from moisture-originated movements and if well attached to the supporting construction. Typically, NRCA recommends mechanically fastened sheet (see discussion in Section 4.2).

The following roof decks require use of the overlay board prior to application of the SPF:

- wood-plank and wood panel decks
- steel
- lightweight insulating concrete
- poured gypsum concrete
- precast gypsum panel
- cement-wood fiber panels
- thermosetting insulating fills (perlite aggregate mixed with hot asphalt binder)

To gain a better understanding of design considerations, the next section reviews a few typical roof deck constructions.

4. USE OF SPF IN NEW ROOF CONSTRUCTION

This section reviews use of SPF in concrete, wood, and steel decks, since these three types of deck are popularly used for new roofing constructions.

4.1 Structural Concrete Decks

All joints between panels and openings that exceed 6 mm (1/4 in) should be grouted or caulked with elastomeric sealant and/or taped. The surface must be clean from oil, grease, and form release agents and must be sufficiently dry (at least 28-day-old concrete). Sometimes, to enhance adhesion, a primer may be recommended. The following primers can be used: wash primer, or chlorinated rubber, neoprene, urethane, and epoxy.

Normally, the SPF roof on a structural concrete deck would be designed based on the flow-through principle using moderately permeable coatings such as acrylic, urethane, and chlorinated synthetic rubbers and no vapor retarder on the deck side.

Note: The structural concrete has a density of minimum 1360 kg/m^3 (85 lb/ft³). Applying the SPF on insulating concrete (with lower density) without underlay board is not recommended.

4.2 Wood Plank and Wood Panel Decks

Design and construction of this roofing system depend on expected moisture performance of the system. If the roof deck is well attached to the supporting construction, and fully protected from moisture ingress, the expectancy of moisture–originated movements of the wood panels is low. One may apply the SPF directly on the wood surface. This may be done after caulking all joints in excess of 6 mm (1/4 in) with a suitable sealant and then priming wood surfaces. The following primers can be used: chlorinated rubber, neoprene, urethane, and epoxy.

Like the concrete deck, the wood surface must be dry. Wood shall not have moisture content in excess of 18 percent by weight when SPF is applied to a deck, which can dry inwards. Maximum moisture content in the range of 8 to 12 percent would be required if a polyethylene vapor retarder was preventing the inward drying of the deck.

Typically, SPF would be directly applied if the roof were in a hot and dry climate. The SPF roof would likely be covered with a permeable coating forming a fully permeable system operating according to the flow-through principles. The same design with permeable coating on the HD-type SPF would also be used if this deck was constructed in cold climate with limited internal moisture loads, e.g., a residential building.

In hot and humid climate, one would follow the NRCA recommendation for a mechanically fastened base sheet or a layer of preformed thermal insulation. The following materials would be suitable: exterior-grade gypsum board, wood fiberboard, perlite board, or polyisocyanurate foam board provided with an appropriate facing material. Depending on the use conditions, with higher humidity in the indoor space, one would likely select a less permeable coating, e.g., butyl, and eliminate moisture sensitive materials from the overlay selection.

From a moisture design viewpoint, the worst case scenario is a cold storage space located in a hot and humid climate. One would require a moisture non-sensitive overlay board such as a high density extruded polystyrene or polyisocyanurate with a fully adhered vapor retarder (e.g., a modified bitumen membrane) to be placed on the overlay board before applying the HD-type SPF. The external coating of choice is the modified asphalt.

4.3 Steel Roof Decks

The cold-rolled steel sheets with ribs to provide strength and rigidity should not be lighter than 22 gauge. Steel decks should have a factory galvanized G-60 or G-90 coating (see NRCA Tech. Bulletin 15-9). They are typically supported by open-web, steel joist framing.

All joints are lapped (joints over 1/4 inch are caulked with a suitable sealant) and fastened. Typically, an overlay board (with sufficient width and thickness to span or fill flutes) is mechanically fastened to the deck. Fasteners should provide an appropriate wind uplift resistance.

The structural and thermal movements normally encountered in the sheet-steel deck require use of an overlay board in this construction. Some architectural and structural metal decks have sufficient stiffness to permit direct application of the HD-type SPF.

4.4 Roof Details

This section provides architectural details to analyze the SPF design requirements rather than to provide information for a designer or contractor. This section is brief, since the SPFD roofing guide and the NRCA roofing manual each provide more than twenty drawings for SPF roofing.

The first two figures show flashing for a deck that is either supported by a separate, steel construction (Figure 46) or supported by the masonry wall (Figure 47). A main difference between these two situations is the degree of the structural movements between the deck and the wall.

Figure 46 shows that a compressible insulation, either high density glass- or mineral fiber insulation (MFI) board, is used to accommodate for movements. The vertical wood member is attached only to the deck. It is allowed to "slide" when the masonry wall undergoes its initial shrinkage (during the masonry drying) and thermal or moisture-originated expansion/contraction movements. The latter movements have either seasonal or daily character. The MFI board is adhered (or mechanically attached) either to the wall or to the wood (one side only).

The vertical wood member is placed in contact with a horizontal wood-nailer, secured to the deck with fasteners, approximately 600 mm (24") on center. Then, a wood cant strip is placed in the corner of these two wood pieces. The wood cant is to provide rigidity to the vertical wood member (structural strength of the wood joint) and to provide gradual reduction of the thickness of the SPF layer when approaching the termination point. In many design details involving the termination of the SPF, one may observe particular care being taken about the last 150 mm (6 in) of the foam. It is usually tapered to about 25 mm- (1 in) thick layer. In typical foam-stop details, the SPF termination may be reinforced with a polyester fabric installed in a third (optional) application of the elastomeric coating. The termination can also be slowly thinned ("feathered" out) on a previously primed surface of the sub-

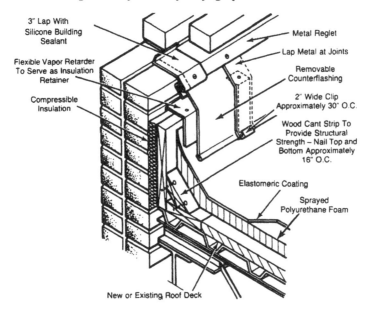

FIGURE 46. Flashing for a non-wall supported deck. Another design, improved with regard to air and moisture control, is shown in Figure 17. Reproduced with permission of SPFD–SPI Inc.

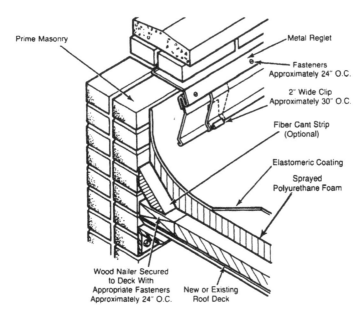

FIGURE 47. Flashing for wall supported deck. Reproduced with permission of SPFD–SPI Inc.

strate. This area is later coated for an extra 100 mm- (4″) distance (Chapter 9 Section 2.2).

The wall-deck connection is simple when no movements are expected in the junction. The only function of the cant profile is to provide gradual reduction of the thickness of the SPF layer when approaching the termination point. The wood cant may be replaced by a fiber cant, typically a MFI board.

The approach to rainwater control is identical in both cases. The counter-flashing is attached to the masonry wall by inserting a metal reglet into one course of the masonry. The strips of metal reglet, attached to the wall with fasteners approximately 600 mm (24″) on center, are covered by 75 mm- (3″) clips. These clips are often attached on a silicone caulking compound (to prevent them from being pushed out of their place during the bricklaying rather than to protect from entry of rain).

The counter-flashing is then inserted into the cleat seam of the metal reglet and permanently fastened (Figure 47) or inserted as a removable counter-flashing (Figure 46). In both cases, the adjacent pieces of flashing are not to be soldered together (breaks in the soldered joints could channel water behind the flashing). Instead, a 50 mm- (2″) clip that is used for mechanical attachment of the counter-flashing to the wall and placed approximately 750 mm (30″) on center can also be used to cover the joints.

Note that the detail shown in Figure 46 often loses the air and moisture tightness within a short period of service. An improved design is shown in Figure 48.

The design shown in Figure 48 differs from the traditional designs where air/moisture control was addressed by means of a "flexible vapor retarder." The WAM-type membrane (see Appendix A6), attached to wall surface and to the primed vertical wood, ensures the continuity of air barrier function. In this design, a high-density MFI board must be attached to the vertical wood surface to slide on the WAM-type membrane that is attached to the surface of masonry wall.

The second line of defense (see Chapter 8) is performed by using either a continuous galvanized steel angle 0.38 mm (0.015″) thick or the WAM-type membrane, attached to the wall, and drawn over the upper surface of the wood member. The galvanized sheet metal expansion cover, placed on WAM-type membrane closure, has traditionally been used in the expansion joints. The counter-flashing shown in Figures 47, 48, and 49, is used primarily for aesthetic purposes.

Figure 49 shows an expansion joint. Expansion joints should be provided when:

- the structural movement is expected
- there is a change in the substrate material
- there is a change in the direction of the roof deck

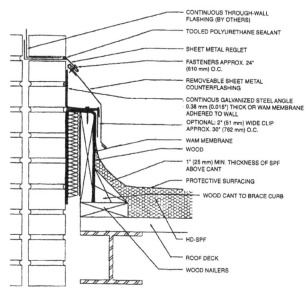

CONTINUOUS THROUGH-WALL
FLASHING (BY OTHERS)

TOOLED POLYURETHANE SEALANT

SHEET METAL REGLET

FASTENERS APPROX. 24"
(610 mm) O.C.

REMOVEABLE SHEET METAL
COUNTERFLASHING

CONTINOUS GALVANIZED STEEL ANGLE
0.38 mm (0.015") THICK OR WAM MEMBRANE
ADHERED TO WALL

OPTIONAL: 2" (51 mm) WIDE CLIP
APPROX. 30" (762 mm) O.C.

WAM MEMBRANE

WOOD

1" (25 mm) MIN. THICKNESS OF SPF
ABOVE CANT

PROTECTIVE SURFACING

WOOD CANT TO BRACE CURB

HD-SPF

ROOF DECK

WOOD NAILERS

FIGURE 48. Flashing for a non-wall supported deck with improved air/moisture control function.

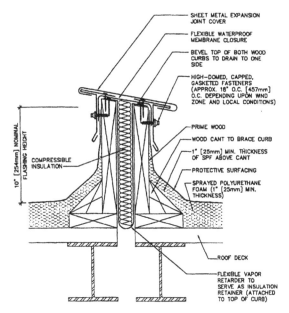

SHEET METAL EXPANSION
JOINT COVER

FLEXIBLE WATERPROOF
MEMBRANE CLOSURE

BEVEL TOP OF BOTH WOOD
CURBS TO DRAIN TO ONE
SIDE

HIGH-DOMED, CAPPED,
GASKETED FASTENERS
(APPROX. 18" O.C. [457mm]
O.C. DEPENDING UPON WIND
ZONE AND LOCAL CONDITIONS)

PRIME WOOD

WOOD CANT TO BRACE CURB

1" [25mm] MIN. THICKNESS
OF SPF ABOVE CANT

PROTECTIVE SURFACING

SPRAYED POLYURETHANE
FOAM (1" [25mm] MIN.
THICKNESS)

10" [25mm] NOMINAL
FLASHING HEIGHT

COMPRESSIBLE
INSULATION

ROOF DECK

FLEXIBLE VAPOR
RETARDER TO
SERVE AS INSULATION
RETAINER (ATTACHED
TO TOP OF CURB)

FIGURE 49. Expansion joint with metal cover. From the NRCA roofing and waterproofing manual, fourth edition. Reproduced with permission of the National Roofing Contractors Association.

Because SPF is much more "elastic" than most construction materials (Chapter 2) it gives some freedom in the spacing of the expansion joints. Often, large roof spans are provided only with area dividers. Since inserting a MFI board between two vertical wood pieces and providing waterproofing with the metal cover makes a small difference in the cost, we would recommend considering "expansion joints" instead of area dividers.

The principles shown in Figures 48 and 49 are identical. However, the flexible vapor retarder, serving as an insulation retainer and attached to the top of the curb, requires a comment. The retarder may be necessary only when the insulation does not have sufficient mechanical resiliency. If a high density MFI board is adhered to the wood (on one side only), the vapor retarder may be eliminated. (Observe that air, vapor, and moisture control is performed by the WAM membrane at the top of the curb and the MFI board is not affected by moisture condensation.)

The top surface of the expansion joint must be sloped. To this end, either a prefabricated cover (often made with neoprene and metal flanges) or sloping wood curbs (Figure 49) are used.

Finally, Figure 50 shows a roof drain. The only design consideration not yet

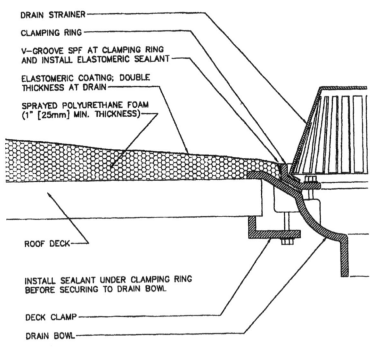

FIGURE 50. A roof drain. From the NRCA roofing and waterproofing manual, fourth edition. Reproduced with permission of NRCA.

discussed is the finish of the SPF layer at the metal edge perimeter flashing, scupper through edge, or clamping ring of the drain (Figure 50). A v-groove is cut in the SPF layer and filled with elastomeric caulking compound.

5. SPF USE IN REMEDIAL ROOFING

A seamless SPF roof may be used in many situations. It provides, however, a special advantage for buildings with unusual shape or configuration and remedial roofing. As the SPF layer may weigh less than gravel removed from the existing built-up roof and as this layer may be built with varying thickness to ensure surface drainage, the SPF roof may eliminate need to tear off existing roofing. This solution, as long as the substrate provides the sufficient resistance to wind uplift, may offer improved performance for a lower cost.

Since a SPF roof offers high resistance to wind uplift (Smith, 1993), an engineering assessment of wind uplift resistance of the existing roof may also be required (with refastening if needed; see recommendations of the single ply roofing institute). Mechanical fasteners provided with 50 mm (2 in) diameter disks can be used to improve adhesion performance of the substrate.

5.1 Presence of Moisture in the Existing Roof

Normally, if the existing roof is wet, it must be dried prior to re-roofing. Both the computer modeling and the experimental results (Desjarlais et al., 1995) show that under specific climatic and use conditions, the SPF roof may be used for re-cover of wet mineral fiber above deck constructions. In the experimental study conducted on a concrete deck, 13 mm (1/2 in) diameter holes were drilled through the deck to provide perforation of vapor retarder. These holes were spaced every 0.6 m (24 in). The re-cover was performed with an average 46-mm (1.8 in) thick layer of SPF and coated with silicone covered with white granules. The actual measurements showed thickness to vary between a 40 mm- and 59 mm-thick layer.

A moisture survey was performed and repeated four, nine, and sixteen months later. Figure 51 shows a comparison between measured and predicted drying of the roof.

The re-cover of the roof gave 10 and 50 percent reduction of the cooling and heating energy requirements (observe that the building is located in Knoxville, Tennessee). It has slab on ground floor construction, not insulated block and brick walls, and windows with single glazing. Energy savings (including those caused by the measured solar reflectance of the white silicone/granules coating) for an assumed 25-year service life and roof re-coating every ten years, showed a saving-to-investment ratio above three. (A

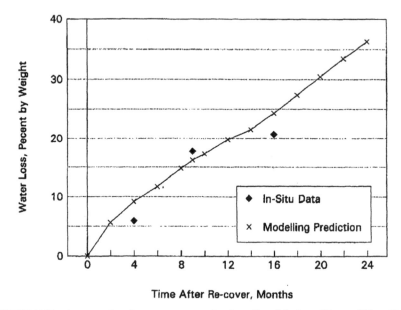

FIGURE 51. A comparison between measured and predicted drying of the roof (from Desjarlais et al., 1995, reprinted with permission).

saving-to-investment of one means that the return on investment is equal to the discount rate.)

5.2 Use of SPF over Existing BUR or Modified Bitumen Membranes

To decide whether to tear off the existing roof or recover it with SPF, one must consider the repair of the structural roof deck, degree of damage of the existing roofing materials and the presence of moisture. If moisture is present in the roof, performing a nuclear moisture survey and evaluation of the roof's drying potential (see previous section) may be advisable.

Most fully-adhered membrane systems may receive a direct application of SPF. The requirement of a clean, dry surface applies to both new and retrofit roofs. Soft mastic, paraffin-containing cold process adhesives and other repair materials, which are known for poor adhesion to SPF, are to be removed.

All loose gravel and deteriorated materials should be removed (using means such as power vacuums or air blower equipment) and should be power washed. The existing roof should be checked for dimensional stability (ridging, splitting, cracking). The existing roof must be inspected for adhesion between plies, insulation, and deck. Blisters, wrinkles, and fishmouths shall be cut out and/or mechanically fastened. Mechanical fasteners provided with 50 mm- (2 in) diameter disks can be used to improve adhesion performance of the substrate. All flashings and gravel stops should be removed or refastened as required. The surface, particularly in the previously coated areas, requires priming that is compatible with both the membrane and SPF. Lightning rods, and electrical or mechanical conduits should be relocated or raised above the finished roof surface prior to the foam application.

If the substrate does not have sufficient slope, the slope must be built up either with the SPF or with other thermal insulation materials that are compatible with SPF. The minimum thickness of the SPF layer is 25 mm (1 in) when applied on rigid boards such as wood fiberboard and 40 mm (1-1/2 in) when applied on less rigid thermal insulating materials such as mineral fiber boards.

If a large area of the membrane was repaired with materials known for poor adhesion to SPF, or if the roof inspection did not ensure adhesion sufficient for wind uplift, the use of an overlay board should be considered. The overlay board would have to be mechanically fastened to the deck.

5.3 SPF over Existing Metal Roof

To decide whether to tear off the existing roof or recover it with SPF, one must consider the repair of the structural roof deck, degree of damage of the existing roofing materials and the presence of moisture. The latter consideration is often the most important since SPF layer applied on the cold side of the wet wood or corrosion-prone metal may accelerate the wood decay or corrosion.

In cold climate considerations, one must bear in mind that metal to metal joints that may be water tight are not really tight for the water vapor transport. Adding SPF will not only increase the resistance for vapor flow through the roof system, but will also increase the metal temperature. Substantial thickness of SPF will reduce the potential for moisture condensation to a minimum. If not enough thermal resistance is added, the condensation still takes place and the risk for corrosion is increased because the vapor flow outward is dramatically reduced. One method of reducing the risk of corrosion is to fill the flutes of the fluted metal deck. Then the SPF is applied on the even surface. In a warm climate, one often stretches a fabric mesh over the deck before applying the SPF layer.

Another important consideration for corrugated and fluted metal decks is the degree of movement. Movement in seams, combined with presence of moisture could cause delamination of SPF from the substrate. Therefore, an overlay board mechanically fastened to the structural deck is a standard solution. Moisture considerations imply that permeable and semi-permeable coatings are used above the SPF layer and products such as an external grade gypsum, wood fiberboard, perlite, high density polyisocyanurate, or polystyrene would be selected for overlay boards.

Most architectural (minimum slope 25 percent) and structural (minimum slope 4 percent) metal roof decks may receive direct application of SPF. All exposed screws and rivets must be checked. When loose, they must be tightened or replaced. All metal sheets and panels must be checked for their attachment to the structural members or deck. When necessary, with a view to wind uplift and fatigue of the existing fasteners, new fasteners must be added.

The cleaning usually requires removing grease and oil with a solution of tri-sodium phosphate and rinsing with fresh water. Soft mastic, paraffin-containing adhesives and other repair materials, which are known for poor adhesion to SPF, are also to be removed. Then the exposed surfaces are to be washed under a pressure of 13,800 kPa (2,000 psi).

Areas of rust should be primed with rust inhibiting primer (e.g., zinc chromate; see NRCA, 1996).

Design and Performance
of Walls with SPF

BUILDING SCIENCE IS on the threshold of a major change, perhaps even a scientific revolution. Incremental changes occurring through many years are now making the impact larger than predicted—not because of any single issue, but because of the change in the paradigm of environmental control of buildings (Chapter 8). The holistic approach (systems approach) is now finding its way from academic circles into building codes and material standards.

The occupant of a building demands that indoor air be fresh, free of pollutants, and maintained within a selected temperature range. To ensure that these requirements are fulfilled, the mechanical engineer must use controlled ventilation (with or without air conditioning systems). To this end, the whole building must be airtight and the indoor space divided into controllable compartments. As a starting point, to provide the airtightness of the building envelope, one may use air barrier systems. The 1995 edition of the Canadian National Building Code requires doing so.

1. DEFINITION OF AIR BARRIER SYSTEMS

The principal function of an air barrier is preventing both the infiltration of the outdoor air into a building and the exfiltration of indoor air to the outside. To this end, the air barrier must meet a number of requirements: air impermeability, material continuity, structural integrity (Lux and Brown, 1989) and durability.

There are, however, different means of "airtightness" control and, therefore, the values identifying these requirements may initially appear confusing. First consideration is given to a complete air barrier system and Table 42 lists the limiting values.

Table 42 implies that in the cold Canadian climate, the amount of moisture carried by air is the factor limiting the rate of airflow through the wall. Indeed, as shown in Figure 52, reporting calculations performed by Ojanen and Kumaran (1996), there is a strict relation between the leakage of the air barrier (AB) system and moisture performance of the wall. Figure 52 shows the maximum air leakage in relation to water vapor permeance of the "external"

187

Table 42. Maximum rate for the complete air barrier system.

Indoor Air, 21°C (70°F), with Relative Humidity	Maximum Flow Rate liters/(m² s)
less than 27	0.15
27 to 55	0.10
more than 55	0.05

part of the wall (i.e., sheathing and siding that does not create accumulation of moisture inside the cavity frame wall).

Table 42 lists maximum flow rates through the whole wall. According to the National Building Code of Canada, testing the whole wall may be avoided if a plane of airtightness (material with sufficient resistance to airflow) is identified and tested for air permeance. The equivalency of the system testing was, therefore, achieved by setting an air permeance requirement of 0.02 l/(m² s) for any material controlling airflow through the building envelope.

The 1995 edition of the National Building Code, in Section 5.4 Air Leakage, defined performance objectives for air barrier systems as follows:

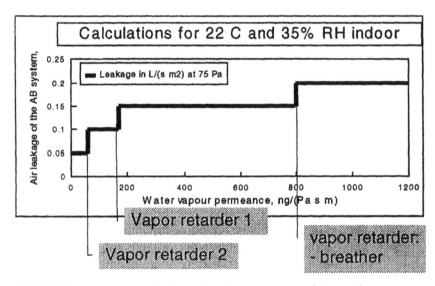

FIGURE 52. A maximum air leakage of air barrier system in relation to the water vapor permeance of the "external" part of the wall (i.e., sheathing and siding that does not create accumulation of moisture inside the cavity frame wall, from Ojanen and Kumaran, 1996).

- Air leakage of the material of the airtightness plane must not exceed 0.02 l/(s m²) @ 75 Pa.
- Barrier continuity over joints, penetrations, etc., must be provided for the whole period of service life.
- Structural design of the system is to be considered for: 100 percent wind loads deflection at 1.5 times wind loads.
- The system components must also conform to their respective material standards.

A SPF wall and two popular systems:

1. Fiberboard sheathing covered with a spunbonded olefin (SBO) and strapping
2. Gypsum with finished joints

were tested for air barrier performance. The SPF was applied to fiberboard sheathing into the wood frame wall (wall 3). The SPF wall was tested with and without taping of the perimeter joints (Brown and Poirier, 1988). Air leakage testing is normally performed in an airtight test chamber to which a tested wall section (e.g., 2.4 m × 2.4 m; i.e., 8 ft × 8 ft) is clamped with an airtight connection. By either pressurizing or evacuating the space behind the specimen while measuring both the flow rate and the pressure difference across the walls, the air barrier performance of the wall can be assessed.

A typical test procedure consisted of two parts: an air leakage characterization, i.e., airflow rate expressed as a function of pressure difference, was determined up to 100 Pa. Then, sustained pressure differences of 250, 500, and 1,000 Pa, simulating static wind loads, were applied for one hour each. Pressure loads 1.5, 2.0, and 2.5 kPa simulating dynamic (gust wind) loads were applied for ten second periods. The same pressure differences were applied in both positive and negative directions. After each change of wind load conditions, the specimen was visually examined for structural damage. Furthermore, the airflow at 75 Pa was measured again and compared with the initial result (Brown and Poirier, 1988).

The wall (1) was constructed as follows. The 11 mm- (7/16 in) thick, asphalt impregnated, fiberboard sheathing was nailed with 45 mm (1 ¾ in) galvanized nails at 152 mm (6 in) distance on the perimeter, and with double spacing along the intermediate supports. Longer, 64 mm (2 ½ in) spiral nails were used at 305 mm (12 in) to fasten wood strapping. At the edges of the test area, 3M tape was used to attach the SBO to the perimeter.

The wall (2) was constructed as follows. The 13 mm- (1/2 in) thick, gypsum boards, were attached horizontally to the wood structure with drywall screws at 203 mm (8 in) at the edge, and 305 mm (12 in) along the intermediate supports. The screw heads were covered with two coats of joint compound and the horizontal joint was covered with two coats of joint compound and tape. Then, the whole surface was painted with two coats of latex paint.

The asphalt impregnated fiberboard sheathing was attached to the wood frame in wall (3) in the same way as in wall (1). The MD-type SPF, with a nominal density of 37 kg/m³ (2.3 lb/ft³) was sprayed by a certified contractor to approximately 76 mm- (3 in) thickness. In some places, particularly along studs, the excess of SPF was cut flush with the face of the stud.

The rate of air leakage obtained at 75 Pa was 0.488 l/(s m²) for the wall (1). The system performed structurally up to 1.5 kPa sustained load test, ballooning when the positive pressure was applied.

The rate of air leakage obtained at 75 Pa was 0.002 l/(s m²) for the wall (2). The system performed structurally up to 1.8 kPa sustained load test, when gypsum boards were pulled off from some screws under the positive pressure test.

During the test of the wall (3), under 500 Pa positive pressure, the fiberboard was pushed off the nails holding it to the studs. This, however, did not affect the airtightness of the wall, and there was no observed change in the flow rate at 75 Pa. The frame wall with SPF fulfilled all the requirements established for testing of the air barrier. The rate of air leakage obtained at 75 Pa was 0.019 l/(s m²) when fiberboard was taped or 0.025 l/(s m²) for the untaped system. The system performed structurally up to the maximum of the gust load tests equal to 2.5 kPa.

Later, it was established that SPF did not adhere to studs in some locations, yet the airtightness test performed before and after the application of the specified loads showed no difference in measured airtightness. Similar results were obtained by Onysko and Jones (1992); see Section 4 in this chapter.

2. MASONRY WALLS

2.1 Infill Cavity Wall in the Steel Load-Bearing Structure

The wall system comprises a brick veneer, a cavity, an SPF insulation, which also functions as the air barrier, a concrete block wall (often called back-up wall) and a suitable interior finish. If an adequate lateral support is provided, the brick veneer can be replaced by a metal, stone, or precast concrete cladding.

The critical considerations of environmental (heat, air, moisture) protection are:

• Provide a space at the top of the block wall to allow the structural deflection, while providing a lateral support suitable to carry the wind and seismic loads (see Figure 53).

- Since the concrete block will shrink both horizontally and vertically, the airtightness at beams, columns and other penetrations must be assured for a long-term period.
- The latter requirement is usually satisfied by use of strips of WAM membrane (see Appendix A6) placed with 125 mm (5 inch) attachment to the smooth surface of the substrate.
- The flashing required to drain the cavity should be adhered to the backup wall as well as to the air barrier. Different methods to design flashing are discussed later.

To illustrate other critical considerations of environmental protection one may examine Figure 53.

When beam deflections are small, the beam has no clips and the cavity between the brick face and the steel beam does not exceed 150 mm (6 inch) one may use the solution (b). The following comments are made in relation to Figure 53.

MD-type SPF: Recommended thickness is 50 mm (2 inch). It provides a nominal thermal resistance of 2.45 $(m^2 K)/W$ or R 14 $(ft^2 hr \,^\circ F)/BTU$ and a long-term thermal resistance of 2.1 $(m^2 K)/W$ or R 12 $(ft^2 hr \,^\circ F)/BTU$. The

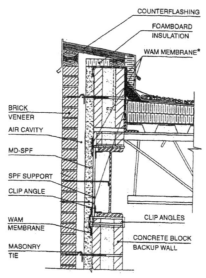

* WAM MEMBRANE = WATER, AIR, MOVEMENT MEMBRANE

FIGURE 53a. Intersection of the masonry cavity wall on steel structure with a roof. This design is based on 1966 SPF Manual by Chem-Thane Engineering Ltd (permission granted).

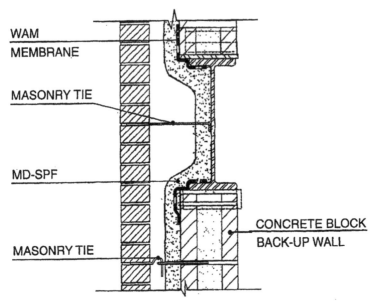

FIGURE 53b. Intersection of the masonry cavity wall on steel structure with a roof. This design is based on 1966 SPF Manual by Chem-Thane Engineering Ltd (permission granted).

MD-type SPF satisfies the air barrier requirements (airflow smaller than 0.02 l/s m² at 75 Pa). When applied on masonry block substrate (see Chapter 2), MD-type SPF also satisfies requirements of vapor retarder 15 ng/Pa s m² (1/4 perm).

Air cavity: A cavity with a nominal 19 mm (3/4 in) thickness is recommended. From the viewpoint of moisture performance, the cavity could be as thin as 6 mm (1/4 in). The cavity adjacent to organic material should not be wider than 25 mm. If a continuous cavity next to the SPF foam does not exceed 25 mm, this prevents oxygen supply in a fire and increases fire safety of the building envelope. Since SPF is not sensitive to moisture, a small cavity with sporadic mortar plugs (mortar falling behind the brick veneer) has no effect on SPF performance. Two reasons, however, support the 19 mm-thick cavity, the ease of foam application (it is not easy to control growth of the foam layer so accurately) and access to electrical wiring.

Counter flashing: A metal cap, nailed at the front face, and sloped to the roof, is necessary. This cap must be treated as counter-flashing, i.e., will never be watertight, and treated (preserved) wood must be used for blocking. The real moisture protection is achieved by the WAM membrane placed over the top of parapet wall and SPF layer.

Insulation: Polystyrene (either extruded or molded product), is recommended here. To be attached to the SPF layer, the insulation should be placed before applying the SPF layer.

Membrane: Either an SBS modified bitumen, peel and stick or thermofusible membrane (Vedagard or Blueskin TG by Bakor Inc., Bituthane 5000 by W.R. Grace & Co., Sopraseal 180HD by Soprema Inc., or equivalent membrane is denoted as WAM (water, air, and movement). WAM membrane is to fulfill requirements specified in Appendix A6.

The WAM membrane will control the air tightness of roof-wall intersection. Both an overlap with roofing membrane and wood-blocks to keep the membrane in place is required. Observe that except for the wood blocking anchored in the masonry parapet wall, there is a continuous insulation on the top and the roof side of the membrane. These precautions were often neglected in older designs. These precautions are, however, necessary to reduce heat flow (thermal bridge) and avoid negative effects of moisture condensation on durability.

Clip angles: Different solutions are used to provide the lateral support for block wall, while leaving a space to accommodate creep of the masonry. Depending on expected deflections of the steel beam and movements of the block wall, this may be achieved by one C-profile (or two clip angles mounted on a steel plate or two clip angles directly attached under the steel beam). Another solution involves two C-profiles sliding into each other, the inner is mounted on the top of the block wall and the outer is attached to the steel beam.

Another clip angle is used to support OSB or plywood, which if placed in the face of the block-wall will provide a substrate for the SPF layer.

Joint support: The same membrane is used to ensure airtightness at the top and bottom of the steel beam. Although SPF is acting as an air barrier, in all locations with movements as well as all expansion joints, a 300 mm (12 inch) strip of the WAM membrane will function as the airtightness plane. This may be considered a second line of defense (Bomberg and Brown, 1993). Without the elastic joint support, as shown here, SPF would have to be tested to prove that cracks would not develop there during mechanical fatigue cycling.

Older designs often used caulking under the steel beam. It is not recommended since this location is not accessible after the construction has been completed. Loss of air tightness of this joint would be difficult to correct.

SPF support: External grade gypsum board may be used for very low beams, normally a minimum 19 mm (3/4 inch) thick OSB or plywood would be mounted on angles and used as the support for the SPF layer. Such a strong support is needed because of the possible warpage during foam curing (a different rate of air ingress).

Comment: A flat roof slope must be at least 1:50 (Brand, 1990, Baker, 1980).

Figure 53 highlights critical considerations of environmental protection, namely:

- A waterproofing membrane links the vertical air barrier/drainage plane over the parapet wall to overlap with the roofing membrane. This approach is recommended in contrast to old-fashioned methods relying on water-tight flashing on the top of the parapet wall (caulking, soldering). Water-tight flashing often became damaged or inefficient with service time.
- A high performance peel and stick or thermofusible membrane fulfilling specific fatigue (movement) requirements is required and denoted as the WAM membrane. This membrane controls both the air tightness and waterproofing of the joint. To ensure its long-term performance, the membrane is protected from damage by a piece of thermal insulation and wood blocking. A metal counter flashing provides a decorative element as well as protects the wood cap needed for achieving a slope on the parapet wall.
- Brick veneer must have a space for expansion at the underside of the shelf angle. It is recommended that a caulking compound be placed on a backer rod (polyethylene rope) and the space behind it not be filled with any compressible filler.

2.2 Details in the Masonry Cavity Wall

Figure 54 shows detail of steel frame in the masonry cavity wall.

Again, as in all other instances, the WAM membrane is used as the second line of defense (Bomberg, Brown, 1993). Without use of this membrane to support SPF at the location of movements, we would require SPF to be tested for the mechanical fatigue and for the resistance to crack propagation under cycling environmental conditions.

The joint in the brick veneer is wider than in the block backup wall. Therefore, a compressible filler is placed behind the backer rod, while the filler is not used in a block wall. The caulking in the block wall is used for aesthetic purposes only (Figure 55).

Figures 56 and 57 present wall-window interface details where the continuity of the SPF air barrier is extended across the wall-window intersection with the use of the WAM membranes placed over the top of the block wall. As in Figure 53, performance of this membrane decides upon the airtightness of this detail. Therefore, wood blocks are used to protect the membrane and keep it in place.

Caulking, shown in these two details, is applied for two different purposes. It may be used as an aesthetic element of the facade finish, e.g., under the edge of flashing. It may also be used for controlling water entry into the assembly, e.g., under window frame. In the latter case, the joint design must follow spe-

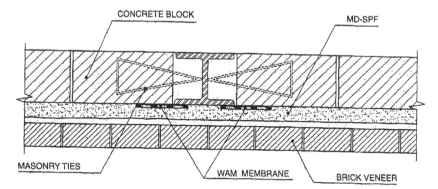

FIGURE 54. Steel frame in the masonry cavity wall. Reprinted with permission from 1996 edition of the SPF Manual by Chem-Thane Engineering Ltd.

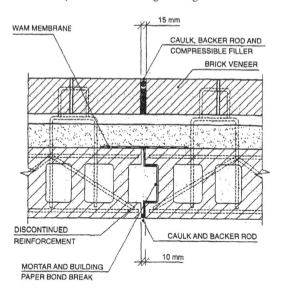

FIGURE 55. Control joints in the masonry cavity wall. Reprinted with permission from 1996 edition of the SPF Manual by Chem-Thane Engineering Ltd.

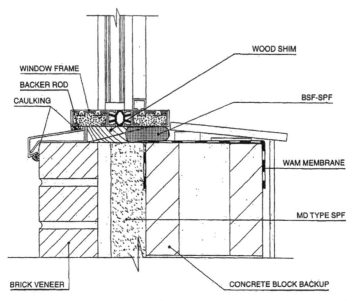

FIGURE 56. Wall-window interface, sill plate. This design is based on 1996 edition of the SPF Manual by Chem-Thane Engineering Ltd. (permission granted).

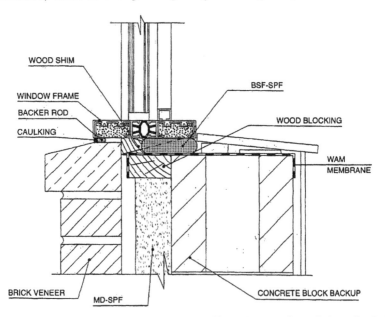

FIGURE 57. Brick veneer with a caping stone wall-window interface, sill plate. This design is based on 1996 edition of the SPF Manual by Chem-Thane Engineering Ltd. (permission granted).

cific requirements of width and thickness. Furthermore, the backer rod must be in place before the caulking is applied.

While airtightness of the wall is achieved by the SPF-membrane combination, the airtightness of the wall-window interface is achieved by use of the BSF-type SPF (bead applied foam sealant could be either one- or two-component foam).

Figures 58 and 59 present another detail of window interface with masonry cavity wall, namely the window head. The continuity of the SPF air barrier is extended across the wall-window intersection with the use of the WAM membrane. In this configuration, however, the WAM membrane also functions as the drainage plane to lead away any water that might have entered the cavity.

There are two ways of attaching WAM membrane to the block surface. Figure 58 shows the WAM membrane placed within the block joint.

A smaller shelf angle used for the support of the brick veneer (the size and location of the angle is selected by the structural engineer) is detailed in Figure 59. Two separate strips of membrane are used in this design, one on each side of the cavity. After the inner membrane strip is applied, a piece of expanded polystyrene[60] is placed. Then, the second strip of membrane is applied over the polystyrene and shelf angle, stretching outside the edge of the latter. It is attached to the masonry substrate on the 125 mm- (5 inch) strip. During the renovation of the smooth masonry surface, use of the priming liquid before membrane installation is recommended for the second method of the attachment (see membrane manufacturer's instructions).

The wood shims used to mount the window frame are protected by the external caulking (applied on the backer rod) and the BSF-type SPF filled from inside and providing the airtightness of the wall-window interface.

Figure 60 shows a durable, high-performance bay window construction. Though the rest of the wall is a masonry cavity wall the window is built in a wood frame construction. The MD-type SPF is applied under the bay window floor and on the outside of the foundation wall header to the same thickness as in the wall above. SPF is applied from outside in the external corners. Then it is shaved to fit under the OSB or plywood sheathing. The MD or LD-type SPF can be used.

Notes MD-type (LD-type) SPF: Recommended thickness is nominal 75 mm (3in). MD-type SPF provides a nominal thermal resistance of either R21 (ft²hr°F)/BTU, 3.67 (m² K)/W, and a LTTP of 18 (ft²hr°F)/BTU, 3.15 (m² K)/W. For OCF-type SPF, nominal and long-term thermal resistance are nearly the same and equal to R10.5 (ft²hr°F)/BTU, 1.84 (m² K)/W. The MD-type SPF satisfies the requirements for 60 ng/(Pa s m²), or 1-perm vapor

---

[60]If the indoor space has humidity, the extruded polystryene would be used.

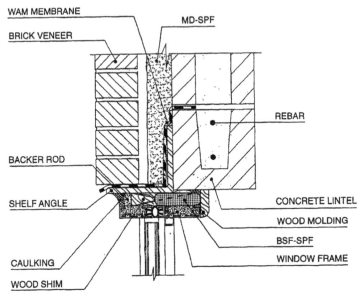

FIGURE 58. Wall–window interface, head and flashing in a detail with large shelf angle. The design is based on the SPF Manual by Chem–Thane Engineering Ltd (permission granted).

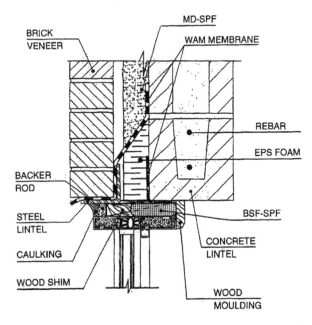

FIGURE 59. Wall–window interface, head and flashing in a detail with small shelf angle. The design is based on the SPF Manual by Chem–Thane Engineering Ltd. (permission granted).

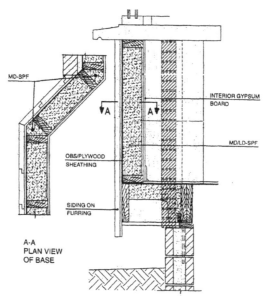

FIGURE 60. Bay window in wood frame that is inserted into the masonry wall. Reprinted with permission from 1996 edition of the SPF Manual by Chem-Thane Engineering Ltd.

retarder. The OCF-SPF will normally require a vapor retarder. For a 2″ × 6″ frame wall, one may use a poured SPF. Since polyethylene (PE) vapor retarder sheet may be punctured and taped after SPF installation, poured OCF may be quite appropriate.

Siding on furring strips: A thin air gap created by the furring strips is very important. This gap permits the use of any type of wood siding without concerns for paint peeling or dimensional instability of the siding caused by entrapped moisture.

OSB/plywood sheathing: This type of sheathing is recommended for use with MD-type SPF, though the exterior grade gypsum may also be used (except for coastal areas with limited drying and hot and humid climates). Using OCF-type SPF with exterior grade gypsum may also require using the furring strips to form an external air gap. Typically the air gap is 12 to 19 mm (1/2 in to 3/4 in) wide and is provided with flashing and drainage capabilities.

Figures 61 and 62 show the connection between the wall and the foundation. Unfilled vertical joints are periodically left in the first layer of the masonry to provide the required drainage. To block the drainage openings and prevent mortar droppings from reaching the bottom of this air gap, coarse aggregate stones may also be placed at the bottom of the cavity. Figures 61 and 62 differ with masonry ties and different means to achieve air tightness and capillary break between foundation and wall. In wood frame wall one may use

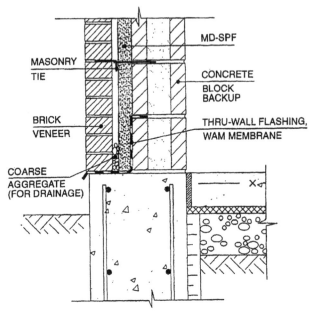

FIGURE 61. Masonry wall at foundation detail. Reprinted with permission from 1996 edition of the SPF Manual by Chem-Thane Engineering Ltd.

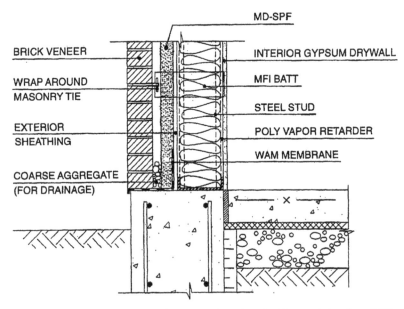

FIGURE 62. Wood frame wall at foundation. Reprinted with permission from 1996 edition of the SPF Manual by Chem-Thane Engineering Ltd.

either compressible gaskets or WAM membranes. Concrete block-wall would require filling of the first course to achieve the tightness. (Often filling of the whole wall or parging on inner surface is necessary to avoid air movements within the wall.)

Finally, Figures 63 and 64 show fire-stop details also developed by Chem-Thane Engineering Ltd. A typical requirement for the horizontal dimension is 20 m (66 ft) and for the vertical dimension is 3 m (10 ft).

Technically, these requirements may be considered as the second line of defense if the concealed air gap is not more than 25 mm (1 in) wide. Detail (A) uses a continuous, galvanized metal track attached to the block wall. A strip of 75 mm (3 in) wide WAM membrane is applied to the primed surface above the metal track. A 100 mm (4 in) wide and 25 percent thicker than the distance to the brick veneer (25 percent compression) strip of high density min-

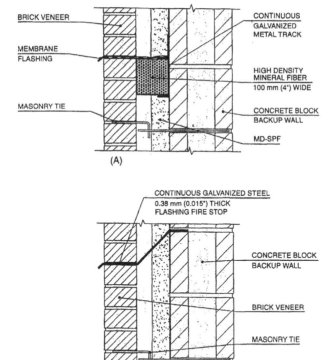

FIGURE 63. Horizontal fire stopping in cavity wall. Reprinted with permission from 1996 edition of the SPF Manual by Chem-Thane Engineering Ltd.

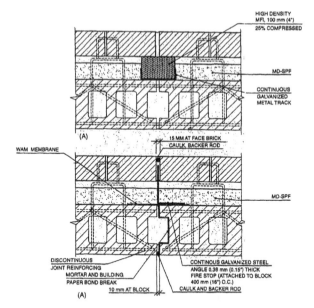

FIGURE 64. Vertical fire stopping in cavity wall. Reprinted with permission from 1996 edition of the SPF Manual by Chem-Thane Engineering Ltd.

eral fiber insulation is inserted in the metal track. (Any glass, slag or mineral product with a minimum 130 kg/m³ (8 lb/ft³) density can be used.) SPF is applied both above and below the metal track to complete the air seal.

Detail (B) uses a through-wall metal flashing with 0.38 mm- (0.015 in) thick galvanized steel. As in detail (A), the SPF, applied both above and below the metal track, is to complete the air seal. A vertical fire stop utilizing the galvanized metal as the fire stop would not have sufficient airtightness (even though it has a caulk on a backer rod). A WAM membrane is, therefore, first applied to the block surface to provide control of airtightness while allowing some movement of the masonry block wall. The galvanized steel angle is then attached to the block (through the membrane). The SPF layer will provide overall airtightness and the WAM membrane will eliminate its cracking in the fire-stopping joints.

3. STEEL FRAME WALLS

The relative difference in stiffness of brick veneers and that of steel studs makes protection against rain penetration much more important in this type of construction. Other requirements remain unchanged. Figure 65 shows an intersection of the brick veneer steel stud (BVSS) wall with the low roof.

Notes: MD-type SPF: Recommended thickness is 38 mm (1-1/2 in). It provides a nominal thermal resistance of 1.84 (m² K)/W or R 10.5 (ft² hr °F)/BTU and a long-term thermal resistance of 1.58 (m²K)/W or R 9 (ft² hr °F)/BTU. The MD-type SPF satisfies the air barrier requirements (airflow smaller than 0.02 l/s m² at 75 Pa). When applied on exterior gypsum sheathing, OSB or plywood substrates (see Chapter 2), MD-type SPF also satisfies requirements of 60 ng/Pa s m² (1 perm) vapor retarder.

Air gap: Typically, in the range between 12 and 19 mm (1/2 in and 3/4 in), see legend to Figure 53.

WAM membrane: Requirements for WAM membranes are listed in Appendix A6.

Exterior gypsum sheathing: The exterior gypsum sheathing may be replaced by 19 mm- (3/4 in) thick OSB or plywood sheathing.

MFI batts: Since one side of the MFI batt is open to air space, workmanship during this MFI installation must be supervised. The design shown in Figure 65, however popular, is often not thermally effective under service conditions. The mineral fiber insulation is facing an open-air cavity, which together with effects of thermal bridging (steel studs) may result in much lower *R*-value than its nominal estimate. To improve thermal performance, one may use "high performance batts" (with higher *R*-value for the same nominal

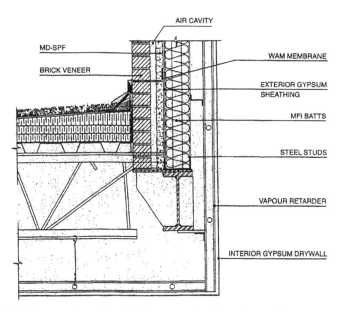

FIGURE 65. Intersection of the brick veneer and steel stud wall with low roof. Reprinted from 1996 edition of the SPF Manual by Chem-Thane Engineering Ltd. (permission granted).

thickness) and place a breather type, vapor-retarder on the open side of the batt. One may also replace MFI with a material less sensitive to effects of air convection (see Figure 66).

Vapor retarder: Polyethylene with thickness of 0.15 mm (6 mil), carefully placed to form a continuous layer, 300 mm- (12 in) overlap and acoustic caulking on joints are recommended.

Locating the SPF layer, which functions as both air barrier and thermal insulation, on the outside (see Figures 65 and 66) reduces the effect of thermal bridging caused by the steel studs. It also provides a good solution for water management and flashing out of the wall cavities.

Figure 66 shows the same detail but is redesigned to avoid possible effects of convection caused by the installation faults if these faults occur at joints and edges of insulation, where thermal bridges already exist. The MFI batt is replaced by a poured LD-type SPF. Since the vapor barrier may be punctured and taped after the application of a poured SPF system, the poured SPF is easy to install.

Notes: MD-type SPF: Recommended thickness is 38 mm (1-1/2 in). It provides a nominal thermal resistance of 1.84 (m² K)/W or R 10.5 (ft² hr °F)/BTU and a long-term thermal resistance of 1.58 (m²K)/W or R 9 (ft² hr °F)/BTU. The MD-type SPF satisfies the air barrier requirements (airflow smaller than 0.02 l/s m² at 75 Pa). When applied on exterior gypsum sheathing, OSB or plywood substrates (see Chapter 2), MD-type SPF also satisfies requirements of 60 ng/Pa s m² (1 perm) vapor retarder.

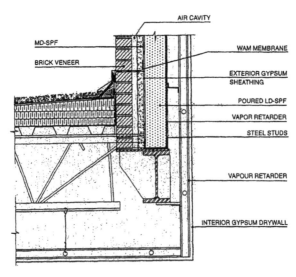

FIGURE 66. Alternative design of the same assembly as shown in Figure 65.

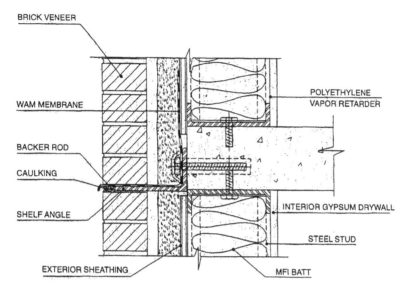

FIGURE 67. Shelf angle supporting brick veneer. Reprinted from 1996 edition of the SPF Manual by Chem-Thane Engineering Ltd. (permission granted).

Air cavity: Typically, the air cavity ranges between 12 mm and 19 mm (1/2 in and 3/4 in); see legend to Figure 53.

Membrane: WAM membrane is defined in Appendix 6.

Cementitious board sheathing: This type of board sheathing is recommended when LD-foam SPF is used for cavity filling.

LD-(OCF)-type SPF: Either a poured or sprayed type can be used as a replacement of MFI batts. The poured type would fill the whole cavity thickness, the spray need not be thicker than 50 mm (2 in). If higher thermal resistance of the wall is needed, the thermal resistance should be increased by placing MD-type SPF outside the stud wall rather than inside the stud wall. An increased thermal resistivity of the cavity insulation makes sense only when steel studs have special thermal breaks to reduce their thermal bridging effect. As a rule, at least 25 percent of thermal insulation should be placed outside the steel studs.

Vapor retarder: A polyethylene vapor retarder (barrier) with thickness of 0.15 mm (6 mil), carefully placed to form a continuous layer (e.g., 300 mm-overlap and acoustic caulking on joints) is recommended for this type of construction.

Figure 67 shows a concrete floor slab which also functions as an airtight space divider. Top and bottom C-profiles are directly attached to the concrete surface.

The space under the shelf angle also allows expansion and shrinkage of the brick veneer while the caulking compound placed on the backer rod provides some resistance to water penetration. Traditional metal flashings are often replaced by reinforced polyethylene membrane or WAM membranes. Observe that to drain water out of the cavity, the flashing must come to the front face of the brick veneer.

4. WOOD FRAME WALLS

This chapter addresses some aspects of thermal, air, and moisture performance of wood frame walls. The systems approach to buildings is the goal of the second part of the monograph. By presenting the best traditional wood frame designs, one highlights critical issues and enhances a comparison with other designs introduced in this monograph.

4.1 Airtightness of the Frame Wall Built with SPF and Wet Lumber

Seldom is the lumber used on a building site a dry, high-quality wood used for framing the walls in the experiments reported in Section 1. (Such wood was selected to improve measurement precision when comparing them with calculations; see Section 4.3.) The current section reports the experiments of Onysko and Jones (1992) with wood framing conditioned to maintain moisture content over the fiber saturation point. One wall was built using 11 mm- (7/16 in) thick wafer board sheathing, and another using standard expanded polystyrene (EPS) sheathing. The objective of this study was to examine whether the construction moisture could dissipate without causing the wood to decay and whether these walls could maintain airtightness over time.

Except for using wet lumber, Onysko and Jones tested according to the same protocol, using the same foam and applicator as reported by Brown and Schwartz (1987). Then the wall was stored and re-tested when reaching the required, lower moisture content.

Drying of SPF insulated walls was compared with a full-length spruce stud of the same size as used in the test walls. Drying the spruce stud from moisture content of 60 percent to fiber saturation (28 percent to 30 percent) took about 3.5 days, and 8.5 days more to dry to a 19% moisture content. Conversely, it took 35 days for walls with SPF to reach 19 percent when the initial moisture content was 25 percent to 33 percent.

Airtightness of both walls was tested only after a period allowing development of sufficient wood shrinkage. Nonuniform shrinkage usually results in warping of the wood and may lead to local failures of the bond between the

SPF and wood. Airtightness tests were therefore performed 42, 65 and 85 days after installation. The fraction of shrinkage was estimated at about 60 percent in the first test and 80 percent at the last test (Onysko and Jones, 1992).

Results indicated that some change in the air leakage took place immediately after the initial wind pressure loading, but that the wall airtightness remained constant for the rest of test period. For polystyrene-clad wall, the leakage prior to the static loading was 0.1 l/(s m²). The air leakage rate increased to 0.124 after the load application. For wall with wafer board sheathing, the change in air leakage caused by the static load was smaller, and the level of air tightness was much higher, 0.041 l/(s m²).

In conclusion, Onysko and Jones (1992) stated that using wet lumber for framing may introduce only a small change in airtightness of the walls. Other concerns, however, must also be addressed as the drying rate of the wall was significantly reduced.

When there is a prevailing thermal gradient and the SPF is applied on the cold side of wet wood, one must expect a long drying time. This was documented in two field failures—in one case, by encapsulating wet wood planks with SPF applied on the cold side (Quebec), in the other case by applying foil faced sheathing and SPF in the wall cavity (northeastern part of the United States). In both cases, the foam insulation on the cold side prevented wet wood from drying and caused wood decay.

4.2 Field Study of Moisture Absorption under Air Leakage Conditions

Four stud spaces in the exterior west-facing wall of an IRC test building were filled with 75 mm thick SPF (Bomberg, 1993). A weather screen was made out of a plywood and hinged at the top to allow easy access and observation of the 13 mm- (1/2 in) thick wafer board sheathing from the outside of the building. On the inside, this wall was covered with 13 mm- (1/2 in) thick gypsum painted with two coats of latex paint. There was no polyethylene film between the gypsum board and the SPF. These four cavity spaces were then subjected to different patterns of air infiltration. Space 1 was left intact as the reference. A number of holes were drilled to simulate different patterns of airflow in the 10 to 20 mm (0.4 in to 0.8 in) thick air space created between the gypsum board and the SPF surface. One 75 mm- (3 in) diameter hole was drilled in the centerline of at 150 mm (6 in) from the top. Two holes were drilled in the space 4, each 150 mm- (6 in) from the end of the cavity.

The interior of the experimental building was maintained at 20°C (68°F) and 40 ± 5% RH from the end of October to the beginning of the following May. A frost buildup was observed under the plywood moisture screen. The frost appeared in roughly oval-shaped patches at the top of the four spaces, and was more pronounced in cavities 2, 3, and 4 than in cavity 1. In the beginning of May, pieces of the insulation with the attached sheathing were cut from the top and bottom of each stud space. No significant moisture was found in any of the foam samples (Schwartz, 1989), even though gamma-ray spectrometer (Kumaran and Bomberg, 1985) and weighing before and after foam drying were performed. The frost patches indicated moisture moving through the SPF, while the lack of significant moisture accumulation at the end of the season indicated that the flow-through principle can be successfully applied to the design of walls with SPF.

A long-term research program was performed at Chalmers University (Ondrus, 1979; Larsson, 1987; Isberg, 1979, 1989; Larsson, 1992). This program involved laboratory and field tests in the experimental test building as well as modeling of thermal performance with inclusion of moisture effects. The conclusions derived from this program (Larsson, 1992) support design with the flow-through principle for walls containing SPF.

4.3 Full-Scale Tests of Thermal Performance of Wood Frame Walls with SPF

Table 43 shows details of construction of four types of residential walls where SPF was used in wood frame cavities.

The test procedure as described in the ASTM C236 Standard was performed at three different temperatures to express thermal resistance in relation to the mean wall temperature. The results are given in Table 44.

Thermal resistance (R-value) of the frame wall was calculated as the average of the parallel[61] and series-parallel[62] models specified by the American Society for Heating Refrigerating Air Conditioning Engineers (ASHRAE). The value of thermal conductivity of SPF used in these calculations was the inverse of average thermal resistivity from nine measurements made on three different SPF products fabricated by three contractors and reported elsewhere (see Bomberg, 1993).

Thermal measurements of walls and materials were performed at different temperatures. To compare them, a computer model of multidirectional heat flow through the wall was used instead of the ASHRAE models. Then, extrapolating the measured dependence of thermal resistance on temperature to 24°C, one may estimate what would be the material properties to give the "measured" thermal resistance of the wall at 24°C. In this fashion, the thermal

---

[61]Assuming strictly one-dimensional heat flow through parallel areas.
[62]Assuming identical surface temperature on each of the parallel areas i.e. equalization in the lateral direction.

Table 43. Details of the tested wood frame walls with SPF
(Brown and Schwartz, 1987).

Spec #	Internal Gypsum	Stud Space* Insulation Class 2, SPF Product	Exterior Sheathing	Exterior Siding (hollow back)
4	12 mm /2"	C, 35 kg/m³ (2.2 lb/ft³)	plywood	aluminum
5	as above	as above	XPS	as above
6	as above	B, 35 kg/m³ (2.2 lb/ft³)	EPS	vinyl
7	as above	C, 28 kg/m³ (1.7 lb/ft³)	XPS	as above

Note: Pine was used for 38 mm × 89 mm (2 × 4 in) studs, placed at 406 mm (16 in) on center. Double top- and single-bottom plates and (half thickness) studs on sides were used.

resistivity of SPF was estimated for the walls 4 through 7 and is shown in Table 45. This table also shows thermal resistivity of polystyrene sheathings that were measured at 24°C and used in the calculations.

To address the issue of workmanship, two SPF products and three contractors were involved in this research series. The walls numbered 4, 5, and 7 were manufactured with one foam system, wall number 6 with another; SPF in walls 4 and 7 was applied by one contractor and in walls 5 and 6 by another. Both contractors were instructed to apply foam to a 75 mm- (3 in) depth, but Table 45 shows that the actual thickness varied. The average thickness obtained with a MD-type SPF, density of 35 kg/m³ (2.2 lb/ft³) SPF was close to the prescribed thickness. The thickness obtained when using LD-type SPF, density of 28 kg/m³ (1.7 lb/ft³), was higher than prescribed. More trimming to the stud thickness was necessary.

One may compare thermal resistivity estimated from *R*-value tests on the

Table 44. Full-scale test of thermal resistance (Brown and Schwartz, 1987).

Test	Hot Surface °C	Cold Surface °C	Thermal (m² K)/W	Resistance (ft²hr°F)/(BTU)
4-A	19.4	−6.8	3.10	17.6
4-B	18.7	−20.9	3.09	17.5
4-C	18.1	−34.2	3.10	17.6
5-A	19.9	−6.9	4.24	24.1
5-B	19.4	−20.8	4.29	24.4
5-C	18.8	−34.4	4.32	24.5
6-A	19.7	−6.9	4.18	23.7
6-B	19.3	−20.8	4.24	24.1
6-C	18.8	−34.4	4.26	24.2
7-A	19.8	−6.9	4.70	26.7
7-B	19.4	−20.8	4.74	26.9
7-C	20.2	−34.5	4.76	27.0

Table 45. Thermal resistivity at the mean temperature of 24°C estimated from the measured thermal resistance of the frame wall (Brown and Schwartz, 1987) and measured for sheathings.

Material	Wall Number	Thickness, mm	Density kg/m³	R-Value (m K)/W	R-Value, (ft²hr°F)/ (BTU in)
SPF "C"	4	71.4	35.0	58.9	8.5
SPF "C"	5	73.6	35.0	58.9	8.5
SPF "B"	6	75.8	36.9	58.9	8.5
SPF "C"	7	85.5	27.6	55.5	8.0
sheathing					
XPS	5,7	37.5		35.4	5.10
EPS	6	37.5		24.5	3.54

frame wall (Table 44) with the values measured on the foam samples. For foam product B, the mean value of thermal resistivity measured on thick SPF specimens was 60.1 (m K)/W, or 8.67 (ft²hr°F)/(BTU in), i.e., about 2% higher than one estimated from the wall test. Measurements performed on 7 mm thick specimens cut after a few days aging gave resistivity about 58.2 (m K)/W, or 8.39 (ft²hr°F)/(BTU in), i.e., about 1.2% lower than one estimated from the wall test.

For product C the HFM measurements performed on thick SPF layer gave a value 3.5% higher than that estimated from the wall measurement. However, the measurements performed on 7 mm thick specimens cut after 70 days aging gave resistivity about 58 (m K)/W, or 8.37 (ft²hr°F)/(BTU in), i.e., 1.6% lower than one estimated from the wall test.

In conclusion, the agreement between thermal resistance measured on wall sections with the prediction based on small-scale measurements performed in a HFM apparatus appears good.

4.4 Full-Scale Tests to Verify LTTR of SPF in Wood Frame Walls

After the testing of initial thermal resistance, described in the previous section, the test walls were lifted with a crane and mounted in a West-facing wall of a building in Ottawa. They were exposed to the conditions typical for an air-conditioned office building, periodically taken down, tested in the environmental chamber facility and returned to the field exposure. Results of the tests on wall assembly are given in Table 46. Thermal resistance determined on these walls will be compared with one predicted from laboratory aging of thin foam layers and simplified models of thermal resistance of the frame walls.

These full-scale measurements performed over the period of 5 years were compared with an average LTTR established for both products B and C (see the previous section). All results were recalculated to the effective thickness[63] of 10 mm (Bomberg, 1990) to compare short- and long-term aging data, Figure 67. As two SPF products (B and C) are shown in Figure 68, the scatter of data is higher. Nevertheless, the slope of these curves appears similar (the same effective diffusion coefficient).

Using the scaling method for nominal thickness of wall 6 (75.8 mm) and reference thickness of 10 mm, the scaling factor becomes 57.5 (Kumaran and Bomberg, 1990). For the period of five years, i.e., 1825 days and the scaling factor of 57.5, the scaled time is 1825/57.5 = 31.8 days. At this time, the aging factor (value of dimensionless thermal resistivity) is 0.745. The aging factor multiplied by the mean initial thermal resistivity of 60.1 (m K)/W yields thermal resistivity of 44.8 (m K)/W, or 6.45 (ft²hr°F)/(BTU in).

As thermal resistance of the wall was measured at mean temperature of $-0.8°C$, the 5-year prediction of wall R-value should also be related to this mean temperature. Using the ratio of 1.071 (Brown and Schwartz, 1987) as the ratio of R-values measured at temperatures of -0.8 and $24°C$, thermal resistivity of the SPF layer is recalculated to mean temperature of $-0.8°C$. It becomes $r = 48.0$ (m K)/W which corresponds to thermal conductivity coefficient $\lambda = 0.0208$ W/(m K).

Thermal resistance of the wall may be calculated as the average of the parallel and series-parallel models of ASHRAE (using m = 1, as described in Chapter 8, Section 3.1.2). Using material data identical to those used in the initial wall R-value calculations but varying the thermal conductivity of SPF to match the measured R-value of the wall, the SPF thermal conductivity coefficient of 0.0203 W/(m K) was obtained. The thermal conductivity coefficient predicted from aging curve, $\lambda = 0.0208$ W/(m K) is about 3.5 percent higher than one calculated from actual wall R-value measurements. This means that the actual aging of the wall under field exposure was slightly

Table 46. Thermal resistance versus exposure time of frame wall number 6.

Test	Mean Temp (°F)	R-value, (m² K)/W (ft²hr°F/BTU)	Mean Temp.°C (°F)	R-value, (m² K)/W (ft²hr°F/BTU)	Mean Temp. °C (°F)	R-Value, (m² K)/W (ft²hr°F/BTU)
Aug. 1987	6.4 (43.5)	4.18 (23.7)	−0.8 (30.6)	4.24 (24.1)	−7.8 (17.6)	4.26 (24.2)
Aug. 1988	6.5 (43.7)	4.08 (23.2)	−0.7 (30.8)	4.14 (23.5)	−7.9 (17.8)	4.17 (23.7)
May 1989	6.4 (43.5)	4.01 (22.8)	−0.8 (30.6)	4.07 (23.1)	−8.0 (18.0)	4.10 (23.3)
Nov. 1992	—	—	−0.8 (30.6)	3.80 (21.6)	—	—

---

[63]Effective thickness is the geometric thickness minus the thickness of the destroyed surface layer.

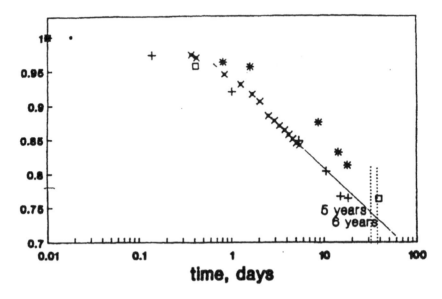

FIGURE 68. Normalized aging curves for SPF products B and C recalculated into 10 mm-thickness (from Bomberg, 1993). Figure 11 from PFCD 1993.

slower than this predicted from the slicing and scaling technique applied to the laboratory exposed thin slices of the SPF.

This agreement obtained between laboratory testing of both thin and thick SPF layers manufactured by three contractors using two commercial SPF products and employing a simplified ASHRAE method to establish the long-term thermal performance of the frame wall is good. It is well within the precision of test methods employed for determination of thermal performance of thermal insulations.

4.5 Full-Scale Tests to Verify Racking Strength of Frame Walls with SPF

Anderson (1993) describes a full-scale study performed on thirty panels with different configurations to verify contribution of SPF to racking (sheer) strength of frame walls with SPF. The racking strength is important as it indicates the resistance of the frame wall to withstand wind loads.

There were twelve 8′ × 8′ (2.4″ m × 2.4″ m) typical wood frame panels with taped and finished drywall on one side and 1/2″ (12 mm) fiberboard or plywood sheathing on the other. There were 18 panels with SPF class 2 and different spacing of studs, namely 16″, 4″ and 32″. The results are shown in Table 47.

Table 47. Measured racking strength (in pounds of force) in relation
to the frame wall and siding type (Anderson, 1993).

Spacing c/c	Vinyl Siding with SPF	Plywood Siding with SPF	Vinyl Siding without SPF	Plywood Siding without SPF
16	2,800	5,300	913	2,800
24	2,420			
32	2,588			
Conventional (with bracing)			3,853	5,262

When compared to the basic walls, with 16″ on center stud spacing but without bracing, one may observe that the presence of SPF increases the racking strength by a factor of two (plywood) or three (vinyl). Since the industry standard today includes bracing and the SPF filled wall without bracing comes close to the same standard, one may either eliminate bracing or use structurally weaker sheathing materials such as fiberboard and achieve satisfactory results.

4.6 Construction Details in Wood Frame Walls

Selection of the wall system is normally driven by two considerations: type of external cladding and the considered level of thermal performance (which inexperienced people usually equate to the required thermal insulation). Following it, consideration is given to the air barrier system and requirements for vapor retarder. Since the frame wall construction is not affected by the selection of the cladding (siding, brick, or stucco), only environmental considerations are discussed below (Table 48).

To benefit the reader, this monograph shows details developed for the tra-

Table 48. Environmental control in analyzed systems of wood frame walls.

Function	ADA	EASE	Standard Wall with SPF	High R-Value Wall with SPF
Air barrier plane	ADA[1]	EASE[2]	SPF	SPF
vapor retarder	foil backed gypsum	polyethylene	SPF	SPF
Thermal insulation	Mineral fiber	Mineral fiber	SPF	SPF
External sheathing	Waferboard or external ins.	Two layers of fiberboard	Asphalt impregnated fiberboard	OSB and SPF

[1]Airtight Drywall Approach (ADA) has been documented in: The Air Drywall Approach (ADA) Construction Manual, Energy, Mines and Resources Canada, August, 1984.
[2]External Air System Envelope (EASE) has been developed to provide mechanical strength to breather type membranes, e.g., spunbonded polyolefin sheets.

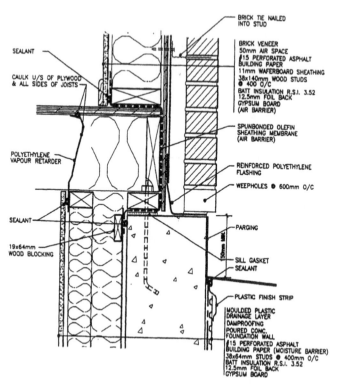

FIGURE 69. A typical frame wall at foundation in the ADA system. From *Best Practice Guide on Wood-Frame Envelopes,* CMHC.[64]

ditional wood frame walls as well as those involving use of the SPF. The ADA (airtight drywall approach) and EASE (external air system envelope) are two classic examples of achieving airtightness in external walls. Two SPF based designs were added, one for a standard thermal performance and one for high *R*-value walls.

The brick veneer wall junction with floor and foundation wall was selected for comparisons. A typical frame wall designed with ADA system is shown in Figure 69.

This detail shows the continuity of the air barrier plane carried from the foil backed interior gypsum around the header of the floor to the foundation wall. The strip of a spunbonded olefin, attached with a bead of a sealant (nor-

[64]Thanks to CMHC's Innovation Centre, project manager Sandra Marshall, to Halsall Associates, and to Otto and Erskine Architects for permission to reprint these details.

mally an acoustic or urethane sealant), goes around the header of the floor to the foundation wall. Another bead of sealant compressed by wood blocking keeps the strip attached to the foundation wall.

The floor joists are laid on the 38 mm × 89 mm (2 in × 4 in) wood sill plate anchored to the concrete foundation wall. The air barrier strip and sill gasket are placed between the sill plate and the concrete wall. Drainage is provided by means of reinforced polyethylene membrane and weep holes.

There is a polyethylene vapor retarder placed in the joist space of the floor. It is caulked and stapled to plywood sub-floor on one side and header of the frame wall of the basement. (If the basement is not finished, two strapping could be used to keep the vapor retarder in place.)

The recommendations for this type of wall include:

- Use special boxes for electrical outlets (provided with internal air barrier gaskets).
- Caulk the wiring penetrations of the exterior wall.
- Create compartments by specifying blocking in the cavity (either compressible gasket or pressure treated wood), in corners, and each 20 ft (6 m) on the longer walls lap and seal the joints in flashing.
- Coordinate installation of air barrier with framers.
- Check that overhang of brick veneer is no more than 1/3 of the brick.
- Check that drywall is compressing the sealant around the perimeter.

Figure 70 shows the ADA system applied to the wall with external insulation at the foundation.

An additional airtightness is obtained by the layer of the external cellular plastic insulation, or in the case shown in Figure 70, by the spunbonded olefin (SBO) membrane located on outside of the mineral fiber insulation. The SBO is taped on all joints.

Figure 71 shows the same detail of the EASE system as Figure 69. This detail shows that the air barrier plane ends at the foundation wall. A strip of the SBO wrapped around the inner fiberboard sheet is overlapped with the main sheet, which is the air tightness plane through the whole height of the building.

The floor joists are laid on the 38 mm × 89 mm (2 in × 4 in) sill plate anchored to the concrete foundation wall and compressing the sill gasket. Care must be taken in selecting and installing this gasket, since this is the weakest link in the control of airtightness of the joint. Other details are similar to the previous design. Drainage is provided by means of reinforced polyethylene membrane and weep holes. The polyethylene vapor retarder placed in the joist space of the floor is caulked and stapled on both sides of the floor.

The recommendations for this type of wall include:

- Continuity of the air barrier at the ceiling level is provided by wrapping the

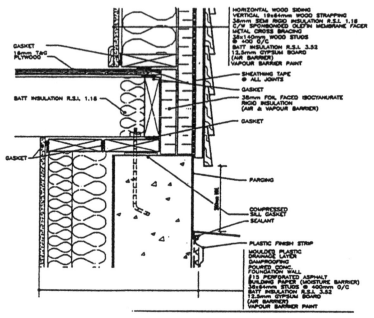

FIGURE 70. A frame wall with external insulation at the foundation in the ADA system. Reprinted with CMHC permission from *Best Practice Guide on Wood-Frame Envelopes.*

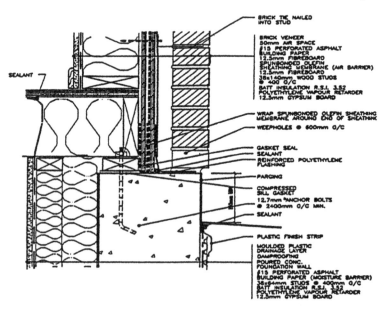

FIGURE 71. The typical brick veneer wall at foundation in the EASE system. Reprint with CMHC permission from *Best Practice Guide on Wood-Frame Envelopes.*

SBO membrane on the top plate and overlapping it with the polyethylene at the ceiling

- Use 1-1/4″ (30 mm) roofing nails for the first layer of fiberboard sheathing and 2-1/2″ (64 mm) long for the second layer.
- Create compartments by specifying blocking in the cavity (either compressible gasket or pressure treated wood) in corners and at each 6 m (20 ft) interval on the longer walls.
- Lap and seal the joints in flashing.
- Check that overhang of brick veneer is no more than 1/3 of the brick.

Figure 72 shows a typical brick veneer wall at foundation using MD-SPF as thermal insulation and air barrier.

The strip of SBO membrane and compressible gasket under sill plate are also used in this design. The compressible gasket is a measure requested in all wall frame designs. If the MD-type SPF was used on both sides of the floor, ensuring the continuity of air barrier, one can use a compressible sheet (foamed elastomeric). This would reduce the flanking noise transmission as well as equalize the contact between wall and floor.

The design shown in Figure 72 uses MD-type SPF for the walls above grade but can use either MD, LD, or OCF types (preferably poured) for the below grade walls. The reasons are as follows:

- Space cost is less in the basement area and using 2″ × 6″ (38 mm × 140 mm) studs, 24″ (610 mm) on center is sufficient to support gypsum boards.
- Pouring SPF through the space created in between floor joists is easy and allows filling of the cavity until overflow comes to the floor joists space.
- If the joist space is also filled with LD foam, a strip of acoustic tile or gypsum board can be used to block the foam overflow in the floor joists space, then the polyethylene strip can be applied.
- If the MD-type of SPF is used for filling the joist space, no vapor retarder is needed.
- There may be a need to cover SPF with a semi-rigid mineral fiber or drywall strips, as all cellular plastics are covered by the thermal barrier. (A limited fire load and difficulty of airflow around the perimeter of the floor joist space makes this requirement more formal than necessary; nevertheless, the interpretation (decision) resides with the local fire marshall).

Figure 73 shows the SPF brick veneer wall at foundation with another design, this time for high R-value performance.

These four examples of wood frame walls with traditional and field-applied SPF in the wall cavity provide a good illustration of how to apply the building science principles presented in Chapter 8. One may observe that use of SPF in frame walls may simplify the construction process while providing a high performance and long-lasting building envelope.

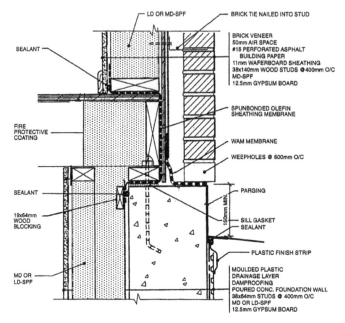

FIGURE 72. A standard SPF wood-frame wall at foundation.

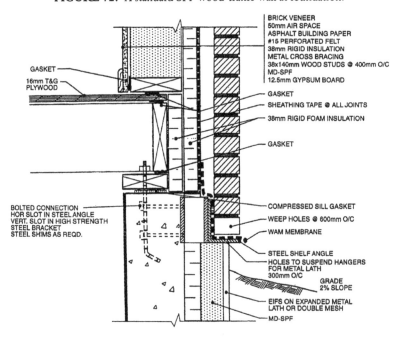

FIGURE 73. The design of high *R*-value SPF brick veneer wall at foundation.

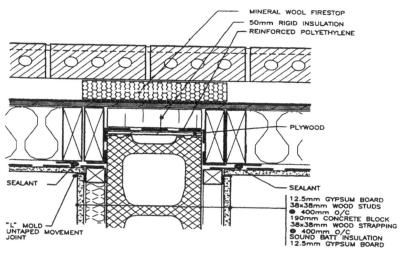

FIGURE 74. A recommended design for fire rated separation of SPF frame walls (party wall). Reprinted with CMHC permission. From *Best Practice Guide on Wood-Frame Envelopes*, CMHC.

To complete this review, Figure 74 shows the design for fire rated separation of SPF frame walls (party wall).

5. METAL BUILDINGS FOR AGRICULTURAL OR INDUSTRIAL USE

SPF has become the insulation of choice for many agricultural structures because it has good long-term thermal performance, is easy to apply in locations otherwise difficult to access, and may be applied on irregular surfaces. SPF may provide an airtight seal and enhance rigidity of the building as well. It is estimated that in Ontario, Canada, at least half of horticultural storage and up to 15 percent of swine barns contain SPF (Muro et al., 1985). In many instances of vegetable storage buildings in Ontario or tobacco curing sheds in South Carolina, the need to provide rigidity to the galvanized metal shell requires direct application of SPF to the metal. This may prevent using a design based on a flow through principle. In such a case, one must decide how much moisture buildup can be tolerated and decide upon additional design measures.

To examine the moisture absorption under such conditions, a long-term experiment (Fraser and Jofriet, 1991) was conducted under controlled conditions. SPF was applied to galvanized steel and acted as a divider between a warm and humid environment (25°C, 90 percent RH) and a cold environment. In the latter, to simulate freeze-thaw conditions, the air temperature was cycled between −8°C and 5°C.

Since the SPF adhered well to the galvanized steel, all moisture moving toward the cold side was trapped and collected in the thin layer of foam-metal interface. The moisture content of this layer was increased with time. After eleven months of testing, nominally 50 mm- (2 in) thick layers of SPF with a density of 39 kg/m³, three samples were removed and their moisture distribution was measured by means of gamma-spectrometer. The zone of wetness stretched between 6 mm and 8 mm from the galvanized steel, and moisture content of this zone ranged from 14 percent to 32 percent by volume. The moisture profile was very steep, falling to as little as 2 percent to 6 percent by volume at a distance of 8 mm to 12 mm from the cold side.

This experiment confirms the test results previously discussed in Chapter 2. In one experiment the flow through was permitted and high moisture content was found only in the middle core of the specimen. In the companion experiment, the impermeable metal facing caused the moisture entrapment at the cold side. The maximum moisture content reached by the polyurethane foam is, however, almost the same in both experiments. Water vapor permeance was to found be independent of moisture content of the polyurethane in the previous study (Schwartz et al., 1989).

In another experiment, the total amount of water trapped in the six tested panels with thickness ranging from 44 mm to 58 mm was found to be roughly proportional to thermal-moisture drive (TMD) (Fraser and Jofriet, 1991). The TMD is expressed as a product of pressure difference, ΔP, and exposure time, t, divided by the foam thickness, L,

$$TMD = (\Delta P \cdot t)/L$$

Effectiveness of an asphaltic coating on wood was checked experimentally (Fraser and Jofriet, 1991). Four combinations were tested, the coating applied either to the wood studs or to the foam, to both of them or to none. When the coating was applied to the wood (with or without application to the foam), the moisture content of the wood remained within 8 percent to 12 percent by weight. When coating was not applied to wood (even the coating was applied to SPF), the moisture content of wood studs increased to 15 percent to 30 percent by weight.

Furthermore, Fraser and Jofriet (1991) examined durability of SPF under freeze-thaw conditions. They proved conclusively that there was no evidence of disintegration of the polyurethane as the result of freeze/thaw cycling performed for 375 days.

These three experimental studies on moisture performance of SPF (Schwartz et al., 1989; Fraser and Jofriet, 1991; Larson, 1992) may assist in de-

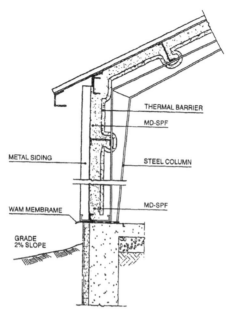

FIGURE 75. Metal building for use with medium or low RH of the indoor air.

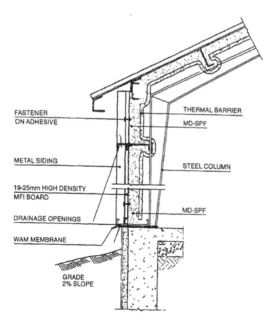

FIGURE 76. Metal building for use with high RH of the indoor air.

signing agricultural buildings with SPF. Such a design should start considering duration of the conditions when the temperature difference consistently
moves moisture in one direction. This period depends on the outdoor climate and indoor conditions (i.e., the intended use). For instance, in the Ontario climate, one may discuss three levels of TMD conditions: light (e.g., ap-
ple storage), medium (e.g., squash and potato storage), and heavy (e.g., stalls
for pigs or hogs).

Consider the construction method that involves building the wood frame,
putting on the outside horizontal steel-cladding directly on the studs, or vertical steel-cladding on strapping. Then, SPF is applied from the inside, directly
on the steel cladding (Fraser and Jofriet, 1991; Hallee and Hunter, 1985). Depending on the level of TMD, different degrees of moisture protection will
be required for this wall. For conditions of small TMD, the only required
moisture protection measure would be the vapor retarder (liquid treatment)
on the inner surface of the wood. For medium TMD, the wood surface must
be treated, and vapor retarding coating (asphalt or butyl) must be placed on
the inner surface of SPF. For heavy TMD conditions, a water impermeable
coating is necessary. For these conditions, a typical use of SPF as directly applied to the metal substrate (Figure 75) can also be changed to include a vapor
pressure equalization/drainage layer (see Figure 76).

CHAPTER 11

Case Studies—SPF Used
for Remedial Measures

THIS CHAPTER REVIEWS a few case studies where a field investigation identifies a performance failure and recommends remedial measures, which include, among others, use of the SPF.

1. CLADDING PROBLEMS IN THE HUMIDIFIED BUILDING

1.1 Description of the Facility and History of Problems

This investigation identified specific problems and recommended remedial measures concerning problems relating to condensation, corrosion,
mold, and other moisture problems at a computer center facility, located in a
cold, heating climate, coded in this report as CC facility.

The condensation of moisture from indoor air deposited water on the back
of the granite panels. The condensed moisture ran down and accumulated in
the "trough" as well as in "gutters" (channels). With insufficient drainage of the
gutters, the condensed water overflowed to the interior and caused damage to
interior surfaces. Furthermore, repeated freeze/thaw cycles created outward
displacement of the granite panels and a hazard of the failure of the granite
cladding.

This facility is a 4 story, steel frame structure with concrete floor slabs cast
over metal decking. The exterior cladding consists of 38 mm- (1.5 in) thick
granite panels connected to steel channels incorporated into a structural
truss system. The granite panels are supported by upturned shelf angles.
Aluminum windows, spandrel glass, and an aluminum bullnose section
complete the cladding surfaces. Internal to the granite panels, associated
support framing, and other cladding elements is a a steel stud wall extending
between floor slabs. This interior steel stud frame wall is insulated with the
faced fiberglass batt insulation. Foil faced gypsum board is installed as the interior surface (foil facing to the cavity side). A 6-mil polyethylene vapor diffusion retarder is installed between the fiberglass and the foil faced gypsum
board (Figure 77).

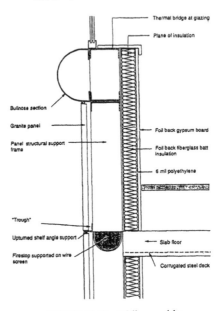

FIGURE 77. Wall assembly.

The granite cladding panel joints were sealed with caulking installed over a backer rod in a "face seal" approach. An interior stainless steel "gutter" was located at each floor to collect condensate from the back surfaces of the granite panels. Intermittent weep-holes were provided at the channel to drain the condensate to the exterior. Visual inspections indicate that many of the weep-holes were clogged with spray-on fireproofing material as well as other construction debris.

The roof consists of a corrugated steel deck insulated with rigid phenolic foam boards 38 mm- (1.5 in) thick, mechanically fastened to the steel deck. An unsupported vapor retarder is installed between the steel deck and rigid phenolic foam board insulation. Fiberboard overlay (3/4 in thick) is placed over the phenolic board insulation. A fully adhered, 5 ply, bituminous BUR membrane is installed over the fiberboard sheathing. Granular material (ballast) is embedded in a flood coat topping to provide mechanical protection and ultraviolet light protection to the BUR membrane (Figure 78).

The space conditioning system consists of 9 air handling units (AHUs) supplying exterior air to the facility (Table 49). Chilled water is supplied to the units for cooling, and a hot water boiler provides hot water for heating.

The only exhaust is from the rest rooms. The exhaust per floor is 750 cfm i.e., 3,000 cfm for the whole facility. In other words, approximately 76,800 cfm more is supplied to the facility than exhausted. On levels 2 and 3 there are

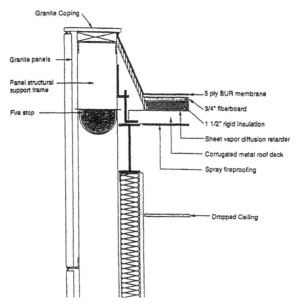

FIGURE 78. Roof/parapet wall assembly intersection.

28 heating and cooling units per floor which maintain conditions at 21°C (70°F), 50 percent relative humidity, year round.

The CC facility, with approximately 400 employees, has operated non-stop since 1991. During the heating seasons, condensation was observed on the interior surfaces of exterior glazing, specifically at window edges. Moisture was also typically observed from the exterior along the edges of the exterior granite panels. During cold weather, water running down the exte-

Table 49. Typical operating conditions for air handling units in the facility.

Unit and Area	Total Supply	Return Air	Outside Air
AHU 1 Level 1	46,000 cfm	22,000 cfm	24,000 cfm
AHU 2 Atrium	24,500 cfm	16,500 cfm	8,000 cfm
AHU 3 Cafeteria	11,800 cfm	none	11,800 cfm
AHU 4 Level 2	12,000 cfm	6,000 cfm	6,000 cfm
AHU 5 Level 2	12,000 cfm	6,000 cfm	6,000 cfm
AHU 6 Level 3	12,000 cfm	6,000 cfm	6,000 cfm
AHU 7 Level 3	12,000 cfm	6,000 cfm	6,000 cfm
AHU 8 Level 4	12,000 cfm	6,000 cfm	6,000 cfm
AHU 9 Level 4	12,000 cfm	6,000 cfm	6,000 cfm
Total Outside Air Supplied			79,800 cfm

rior panel surfaces forming icicles on interior and exterior surfaces was observed.

Water was observed to run out of ceiling cavities at the intersection of ceilings and exterior walls during cold weather. Water was also observed to run out of window head locations. (In an attempt to control condensation, radiant heat panels were installed to raise the temperature of window surfaces.) Water damage included peeling paint at many surfaces. Odors and visible mold growth were noted behind furnishings.

Access openings cut through the interior gypsum board into the exterior wall cavities during the heating season have revealed significant moisture accumulation at the back surfaces of the exterior granite cladding panels. Water and ice have been observed to accumulate in the bullnose section as well as in the "trough" created by the upturned shelf angle supporting the granite panels and the back surface of the granite panels.

During the winter, a serious split in the roof membrane occurred causing emergency repairs. During these roof repairs, significant corrosion of the metal fasteners and metal roof deck were also observed.

1.2 Investigation and Testing

The existing conditions, as-built construction of the building, foundation, wall construction, cladding, roof assembly, and mechanical systems, were evaluated to determine the extent and cause of any deficiencies. Then by conducting a thorough field investigation, see Table 50, problem areas and their causes were identified.

Table 50. The areas of specific investigation in the computer center facility.

Subject of Non-Intrusive Examination	Follow-up
Air pressure profiles under operating conditions	Their relationships
Interior environmental conditions (temperature, RH)	
Environmental service conditions (site location and exposure, seasonal effects)	(Source conditions that impact performance)
Operational and housekeeping practices	
Control of air, water vapor and rain leakage	Their relationships
Window details and window systems	
Roofing details and the roofing system	
The location in the envelope cross sections where dew point (condensation) would likely occur	Review of thermal gradients
Source strength assessment including bioaerosols	
Biological growth (interstitial and on surfaces)	Micro environment

Table 51. Air pressure across the wall assembly.

Location	Total Pressure Drop	On Gypsum Board	On Polyethlene	On Granite
West Level 4	12.3 Pa	0.3 Pa (2.5 %)	1.4 Pa (11.4 %)	10.6 Pa (86.1 %)
South Level 4	10.0 Pa	0.5 Pa (5.0 %)	1.3 Pa (13.0 %)	8.7 Pa (87.0 %)
North Level 2	11.5 Pa	1.2 Pa (10.4 %)	4.0 Pa (34.7 %)	6.3 Pa (54.9 %)
West Level 3	17.0 Pa	1.9 Pa (11.2 %)	2.4 Pa (14.1 %)	12.7 Pa (74.7 %)

Finally, a mock-up test section was prepared at the site and the recommended corrective measures were tested.

Testing of rain penetration, under air pressure difference, was also conducted. The facility owner also provided infrared photographs of the exterior cladding taken during the heating season and provided a detailed review of the cladding condition.

The evaluation was separated into three areas: building envelope, HVAC system, operation and maintenance.

1.2.1 AIR PRESSURE DISTRIBUTION ACROSS THE WALL ASSEMBLY

Air pressure taps were installed across the wall assembly, namely, at the interior, between the interior gypsum board and polyethylene, in the wall cavity, and at the exterior. The ratio of the air pressure drops across the wall components to the total pressure drop across the entire assembly was determined. The individual ratios of air pressure drops are directly related to the tightness of a specific component (element) of the assembly. The tighter an element, the greater the air-pressure drop across that particular element. Conversely, the leakier the element, the smaller the air pressure drop. For example, it would be expected that the air retarder (plane of air barrier system) would constitute the tightest element of the wall assembly and therefore provide the largest air pressure drop. Typically, the air retarder provides at least 75 percent of the total air pressure drop across an assembly.

In the analyzed facility, the existing HVAC system was pressurizing the enclosure 10 to 12 Pascals relative to the exterior under normal operating conditions. Accordingly, creating an additional air pressure difference was not necessary.

Table 51 lists the results obtained at four locations.

The results clearly indicate the absence of the air barrier system. In fact, the results show that the granite cladding panels act as the principal element controlling airflow through the assembly, taking the majority of the air pressure drop.

The investigation proceeded to sealing the openings at the failed caulk joints. It, however, had an insignificant effect on the pressure distribution across the assembly. To confirm it, the investigation also included adding new openings. Again, no measurable change was obtained.

1.2.2 AIR PRESSURE DISTRIBUTION ACROSS THE ROOF/PARAPET ASSEMBLY

At one location of the roof/parapet assembly, the air pressure was determined at the interior, in the parapet cavity assembly, in the roof insulation cavity, at the exterior.

The air pressure drop across the entire roof assembly was 9.5 Pa. The pressure difference between the cavity of the parapet assembly and the exterior was 7.4 Pa and across the roof-membrane was 3.8 Pa. As their sum exceeded 100 percent, it was clear that the parapet cavity was connected to the indoor space. Similarly, the roof insulation cavity was also connected to the interior conditioned space. The latter was confirmed by the review of the assembly details.

Examination of the air pressure relationships showed that the air pressure is higher in the parapet cavity than in the roof insulation cavity. Since they were connected, the air pressure differences create airflow from the interior conditioned space into the parapet cavity, and from the parapet cavity into the roof insulation cavity.

1.2.3 INTRUSIVE DISASSEMBLY

Access openings were cut into the exterior wall from the interior at six locations: North level 2 (two locations), West level 3 (two locations), East and South level 3. Interior gypsum board, polyethylene, and fiberglass insulation were removed to allow the visual inspection. During these inspections, weep openings were often found blocked with fire proofing material. In addition, the back, cavity side, surfaces of the granite panels showed evidence of water stains consistent with condensation of interior moisture.

Thermal insulation was found missing at glazing/wall assembly interfaces (due to the nature of the construction detail). The glazing unit was located at a vertical plane flush with the exterior of the wall assembly, whereas the vertical plane of thermal insulation within the steel stud framing was located to the interior of the wall assembly (Figure 78).

Dropped ceiling panels were removed to allow visual inspection of the underside of the metal roof deck at the upper floor. At numerous locations, fireproofing material was observed to have detached and fallen from the roof deck onto the upper surfaces of the dropped ceiling panels. Water markings and water stains were observed where fireproofing material was observed to have been detached.

The nature of the water damage was consistent with condensation of interior moisture. Indications of substantial and persistent condensation of interior moisture were also observed. Discussions with building staff indicated that some ceiling tiles collapsed from the weight of water and fire proofing.

Large gaps, several inches wide, were observed between the underside of perimeter structural beams and the top of interior gypsum board. No seal or attempt at continuity between the gypsum board and the structural beams was observed. Similarly, no seal or attempt to provide the continuity between the structural beams and sheet polyethylene or foil backing of the fiberglass batt insulation was observed. Inspection revealed that sealing of polyethylene vapor retarder to the structural beams could not be achieved because of the application of fireproofing material to encapsulate the steel beams. A similar lack of continuity was observed at perimeter structural columns and at the intersection of interior partition walls and the perimeter exterior walls.

Air leakage pathways were observed at the intersection of the fluted/corrugated steel deck and the perimeter gypsum board at each floor. Fiberglass batt insulation was installed in the flute areas. Dust markings in the insulation indicate that the insulation at the flute locations is performing as an air filter rather than an air seal.

1.2.4 RAIN PENETRATION TESTING

Rain penetration was tested with an adjustable water spray rack. Air pressure differences across the exterior cladding were induced with the portable pressurization/depressurization equipment. Testing was conducted at one location. A window/bullnose/granite section on the north elevation of level 2 was selected for the testing. The water spray rack was positioned over an area of cladding appearing free from sealant defects.

The water spray rack deposits a uniform film of water over a wall surface to simulate the effect of rain. Water pressures within the spray rack are limited to 30 psi following ASTM standards (E 1105 - 90) which apply to window, curtain wall and precast assemblies. Nozzle design and placement are similarly constrained.

An air pressure difference was also applied across the cladding to simulate the effect of wind. A negative air pressure difference of 75 Pa was applied across the exterior wall assembly. The 75 Pascal air pressure difference is lower than those required by the existing test standards and was, therefore, judged as conservative.

The combination of a water spray rack and a controlled air pressure difference was designed to simulate the effect of wind driven rain on the exterior wall assemblies. After 15 minutes of wetting under controlled conditions, the interior of the wall assembly was carefully examined. No evidence of water entry was found.

Due to the limitations of access, additional testing of the cladding system was not conducted. However, an external visual survey of sealant joints was conducted from the ground with telephoto lens. Numerous breaks in the sealant joints were observed. This was further substantiated by cladding surveys conducted by others.

1.2.5 AIR PRESSURE PROFILES AND RELATIONSHIPS UNDER OPERATING CONDITIONS

A detailed air pressure mapping of the facility was conducted using digital micro-manometers. Zonal and interstitial air pressure measurements were made relative to the central atrium. The air pressure relationship of the central atrium to the exterior was subsequently determined.

The entire facility was found to be operating at between 7 Pa and 12 Pa positive air pressure with respect to the exterior. Levels 2, 3 and 4 were operating at a positive air pressure to the atrium and the atrium was operating at a positive air pressure with respect to level 1.

The air pressure relationships create an interzonal flow path, which moves air out of the upper three levels of the facility, into the atrium, and subsequently into the lower level of the facility.

1.2.6 TEMPERATURE AND INTERIOR ENVIRONMENTAL CONDITIONS

A detailed survey of air temperatures and relative humidity was conducted using hand held digital thermometers and relative humidity sensors. The temperature ranged between 20.5°C and 25°C (69°F and 77°F) and between 48 percent and 52 percent relative humidity throughout the facility. The average temperature was 24°C (74°F) and the average relative humidity was 51 percent.

1.3 Operation and Maintenance of the Facility

1.3.1 DISCUSSIONS WITH FACILITY STAFF

AHU 3, which supplies the cafeteria, operates during the daytime only. This represents an operating time of approximately 60 percent. In other words, 11,800 cfm of outdoor air is intermittently supplied to the cafeteria 60 percent of the time. Air pressure measurements indicated that this affects the facility pressurization by approximately 3 Pa to 4 Pa.

Under extreme cold weather, the percentage of outdoor air is reduced in order to maintain thermal comfort. The outdoor air supply to levels 2, 3 and 4 is reduced by approximately 75 percent (27,000 cfm) to approximately 9,000 cfm (3,000 cfm per floor). Facility staff estimates that these reduced outdoor air supply conditions occur approximately 30 percent of the year.

Under typical operating conditions, the building is pressurized with between 65,000 cfm and 76,800 cfm of supply air in excess of exhaust. Under extreme cold weather, the building is pressurized with between 35,000 cfm and 46,800 cfm of supply air in excess of exhaust. The variation in flow is determined by the intermittent operation of the cafeteria AHU. Under minimal outdoor air supply conditions (extreme cold weather) and with the cafeteria AHU not operating, it is estimated that the facility remains pressurized approximately 4 Pa relative to the exterior.

1.4 Discussion on Failure Mechanisms

The design and operation of the heating, ventilating and air conditioning systems (HVAC) significantly affect the design, construction and performance of the building envelope (the walls, roof and foundation). Therefore, the evaluation of a facility should be considered in terms of the building envelope, the facility's heating, ventilating and air conditioning (HVAC) system, and the facility's operation, maintenance and housekeeping protocols. In particular, relationships between the building envelope and the mechanical systems must be well understood.

The cladding and roofing failures were both caused by the exfiltration of interior, moisture-laden air (Ojanen and Kumaran, 1992, 1995). The lack of an appropriate air barrier system (Di Lenardo et al., 1995) was coupled with the lack of a second line of defense in the design of environmental controls (Bomberg and Brown, 1993).

Ultimately, the granite panels acted as the air barrier and controlled the airflow through the assembly. The granite panels also created the vapor retarder on wrong side of the wall assembly. Unfortunately, the air barrier became condensing surface when warm, interior and humidified air accessed their back surfaces.

Interior, humidified, moisture-laden air was pushed into the exterior wall cavities in effect of the pressurization of the facility and the lack of the air barrier system. Then it condensed on the back of the granite panels. The condensed water was accumulated in the "trough" created by the upturned shelf angle supporting the granite panels and in the "gutters" provided there to collect water. No weep holes were provided, however, at the upturned shelf angles, and many weep holes in gutters were clogged. The condensed water flowed to the interior and caused significant damage to interior surfaces. During winter, the accumulated moisture in the "troughs" and gutters froze.

Repeated freeze/thaw cycles created expansion forces, leading to outward displacement of the granite panels and hazardous conditions. The failure of the granite cladding could occur at any time. Numerous cracks in the panels have been observed during an extensive survey of the exterior cladding independently performed by the facility owner.

Moisture that accumulated behind the granite panels was not able to dry either to the interior or to the exterior air. The "face seal" approach used to seal granite panel joints and impermeable stone made it impossible to dry outwards. The foil backed gypsum board and 6-mil polyethylene vapor barrier produced effective means to prevent the inward movement vapor. Thus, while the moisture was carried outward with air moving through holes and imperfection in details of the wall assembly, this moisture could not return by other means, even when the environmental conditions would permit so.

The roof parapet assemblies were found to be directly connected to both the interior conditioned space and the roof assembly. Interior, humidified air was able to flow laterally/horizontally in the insulation space between the roof membrane and the metal roof decking into and out of the parapet assemblies. An air circulation loop was also found connecting the insulation space under the roof membrane to the roof parapet and the interior conditioned space. Humidified air entering the roof insulation (particularly as the phenolic foam is "hygroscopic") led to corrosion of metal fasteners responsible for transferring the thermal movement and membrane stresses to the structural deck. Once a sufficient number of metal fasteners failed, the stress concentration led to a failure of the whole membrane (splitting).

As in the case of the exterior walls, the air barrier was the exterior surfaces of the roof/parapet assembly. The connection to the wall assembly gave access of the interior, humidified air to the phenolic insulation and fiberboard sandwiched between the roof membrane and the corrugated steel deck. The corrugations created additional air channels that facilitated lateral movement of air in the roof assembly above the steel deck. As the vapor retarder is (probably) unsupported at the corrugations and unsealed at the overlaps the air may access the insulation joints, the cold side fiberboard and even the BUR membrane itself.

1.5 Recommendations

Indoor air in the CC facility is required to have temperature of 21°C (70°F) and 50 percent RH and a positive pressure (relative to the exterior air). These requirements need a specific design of the building envelope (see Chapter 8). Here, the issue is an improvement of the existing building to avoid any future problems.

Firstly, a significant increase in the weep openings on the exterior granite cladding is needed. The weep openings would create a vented (and if possible, even a pressure equalized) cavity located to the exterior of the air barrier. The weep opening geometry needs to be designed in such a manner as to provide:

- prevention of entry of rain
- drainage of water from behind the granite panels
- sufficient vent area to provide air pressure equalization (provided that a retrofitted air barrier system is implemented)

This vented cavity would allow both the rain and the condensed moisture (if moisture laden air has passed the air barrier) to drain and dry to the exterior. Additionally, both vertical and horizontal compartmentalization of the air space behind the cladding would be required. With the appropriate design, it could be provided vertically at relieving angles and horizontally at glazing openings.

As mentioned above, the key to moisture control of wall assembly is the air barrier system. The existing sheet polyethylene and aluminum foil backing on the fiberglass cavity insulation cannot be made continuous enough to act as an effective plenum of the air barrier system. It is due to the complexity of the existing structural system and intersecting demising walls. Furthermore, these materials lack the structural support to carry the loads.

A preferred location of the air barrier system is on the warm side of the assembly. Typically, the air barrier system provides air-pressure drop of about 75 percent of the total and has air tightness about 10 times higher than the cladding system. As previously discussed, the actual requirements depend on the water vapor permeance and the air tightness.)

Figure 79 shows design for rehabilitation of the wall assembly that involves the following:

- New vent/drainage openings cut into the vertical panel caulk joints and drilled into the bullnose sections.
- Existing gypsum board, fiberglass batt insulation and 6 mil polyethylene are removed.
- Spacer discs are adhered to granite panels from inside, to provide support for rigid fiberglass drainage board.
- Rigid fiberglass drainage board is installed from interior to provide external drainage and support for the MD-type SPF insulation. SPF performs the role of an air barrier, thermal insulation and vapor retarder.
- Foil-backed gypsum board was also replaced. (Foil acts as additional vapor retarder.)

The redesigned wall assembly addresses also the issue of surface condensation caused by thermal bridging at glazing edges. The lack of effective thermal insulation under the metal profile of the bullnose section extending under the sill plate caused the thermal bridge reducing the surface temperature at the glazing edge. A mock up section of the wall was prepared at the site and tested to determine appropriate vent/drainage opening geometries and net free vent areas and the air barrier system tightness. This was done by a series of trials. Opening sizes were increased stepwise, until both requirements were met: the 75 percent of the total air pressure drop took place on the air barrier alone and the 10 to 1 tightness ratio between a cladding and the air barrier were met. The satisfactory openings at the top and bottom of each panel joint, were found to be 50 mm (2 in) long.

Figure 80 shows a redesigned roof/parapet wall intersection. The measures recommended for the roof assembly are as follows:

- Remove existing BUR, fiberboard, insulation, and damaged steel roofing.
- Adhere gypsum board layer as support for air barrier system.
- Install WAM membrane over the gypsum board, at perimeter, and seal it to granite panels with MD-type, SPF to provide continuity of air barrier between wall and roof.
- Adhere rigid insulation over WAM membrane.
- Install 3/4 inch polyproplene mesh ("enka-drain") vent membrane.
- Install fiberboard sheathing over venting membrane. Mechanically fasten fiberboard overlay (through rigid insulation) directly into steel deck.

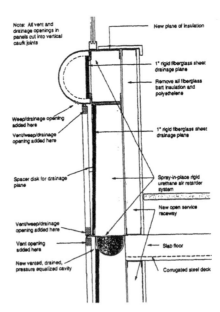

FIGURE 79. Redesigned wall assembly.

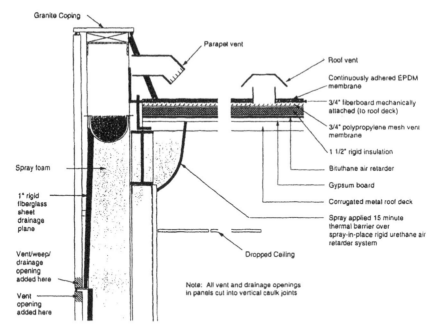

FIGURE 80. Redesigned roof/parapet wall assembly intersection.

- Install continuously adhered, EPDM membrane over fiberboard.
- Install perimeter vents at parapet enclosure, venting vent membrane to the free air.
- Install additional vents through membrane, one vent per 1000 square feet.

2. SURFACE MOLD AND ODORS IN GARDEN SUITES

2.1 Description of the Facility and History of Problems

Some of the 400 units in the residential condominium project constructed in various phases over a period of 1988 – 1993 in Virginia, were experiencing musty odors and mold growth. The units are of wood frame construction. Air conditioning ducts are typically located in vented crawl spaces and vented attics. Foundations are either vented crawl spaces or concrete slabs. Most of the concrete slabs are located in "garden level" units (several feet below grade).

Complaints of musty odors have occurred since their time of initial construction. Reports of surface mold have only occurred in garden level units constructed in the last three years. This appears to coincide with the use of the

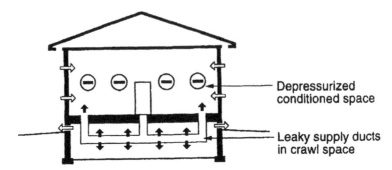

FIGURE 81. Leaky ducts in the crawl space caused depresurization in the house space.

garden level units as rental property during the summer months (previously, most of them remained unoccupied in this period).

During the inspection, particular attention was focused on party wall construction, floor framing, service penetrations, and mechanical system installation. Duct leakage/pressurization testing of one air distribution system was conducted while air pressure fields under typical operating conditions were recorded in several units.

The duct pressurization tests indicated a substantial leakage of air out of the ducts and into the crawl space of the unit tested (Figure 81). The measurements of air pressures indicated significant changes in the relationships within the air-conditioned spaces when air handler fans operated. With the air distribution fans operating, the entire indoor space appeared to have a negative pressure of approximately 3 Pa in relation to the exterior air. With bedroom doors closed, the bedroom air pressure was positive relative to the common areas (approximately 10 Pa difference), while the common areas went to an even greater negative pressure of approximately 5 Pa relative to the exterior.

In one unit, where an air handler was installed in a dropped ceiling location, cycling of the air handler fan led to significant negative air pressures within the interstitial spaces of partition walls. Closer examination of the unit indicated a gypsum board plenum return design was utilized. Gaps between gypsum boards were observed.

2.2 Analysis and Discussion

It has been concluded, based on inspections, field testing, review of construction drawings and discussion with facility staff, that the existing units are experiencing musty odors and mold growth for the following reasons:

- During air conditioning periods, warm exterior moisture laden air infiltrates into indoor space, mostly to bedrooms (leakage of duct work, plenum leakage and negative air pressures relative to the exterior, particularly when bedroom doors are closed).

- Odor saturated air out of crawl spaces infiltrates into indoor space. It is carried through the indoor space and the interstitial cavities because of leakage of duct work, plenum leakage and door closure effects (negative air pressures relative to the exterior).
- Exterior, moisture-laden air from open windows accesses cooled surfaces in the "garden level" units.

The measurement of large positive air pressure differences of 10 Pa, when the interior doors were closed, is illustrated in Figure 82. Air supplied to the bedrooms is unable to return to the air handler due to the lack of a pathway. This leads to the pressurization of bedrooms and the depressurization of the common areas.

The return portion of the air handling system is connected to interstitial spaces. As it leaks, air is drawn out of these interstitial spaces. This leads to a negative air pressure within the interstitial spaces. For example, a gypsum-board return-plenum, with holes, can lead to the depressurization of wall cavities connected by these holes to the plenum (Figure 83). This is consistent with the unit examined with a leaky plenum return.

The air pressure differences created by duct leakage, plenum leakage and door closure lead to the infiltration of exterior air into the depressurized zones. If this air is drawn from the crawlspace, the odors generated there will also be carried into the conditioned spaces with the infiltrating air.

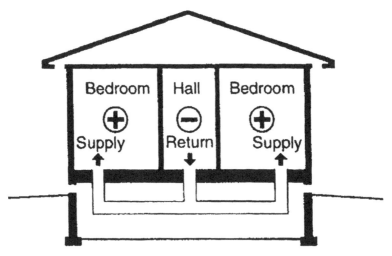

FIGURE 82. In effect of interior door closure, air supplied to the bedrooms is unable to return causing air pressure differences of 10 Pascal.

The complaints of surface mold coincide with a change in operation pattern caused by occupancy of the garden level units. Occupancy brought open windows and sliding doors and increased the humidity of indoor air. Mold growth is directly related to high relative humidity of air adjacent to the wall surface. Where the surface is cooled, the relative humidity of air rises. If relative humidity of such air layer exceeds 70 percent or 80 percent over a prolonged period, mold growth can occur (Hens, 1992).

When warm, humid exterior air enters the air-conditioned space, moisture can be deposited in the areas far away from the entry point (Lstiburek, 1992). For example, if a window is left open during hot, humid weather, exterior air will enter through this window. This moisture may be adsorbed on the surface of textiles, furnishings or carried into the interconnected indoor spaces. This involves multiple and mostly hidden leakage-paths, under base-boards, through electrical outlets or framing details. When warm, humid air enters in the "hollow" wood frame wall and is cooled by the surface of concrete foundation, moisture can be deposited on this surface. This moisture may also cause an increase of the relative humidity in the still air layer at the wall surface, which prolongs the condition of mold growth. This is likely occurring in the "garden level" units, where a substantial part of perimeter walls is placed below grade and therefore is cooler than the others walls. Figure 84 is a detail taken from the construction drawings for the "garden level" units. It shows a wood frame wall constructed to the interior of the concrete foundation assembly.

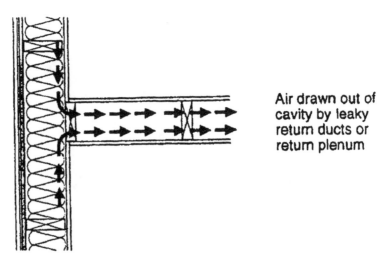

Air drawn out of cavity by leaky return ducts or return plenum

FIGURE 83. Moist, outside air is drawn through the exterior wall and the connected interior partition because of the negative pressure created by the leaky return ducts and the leaky

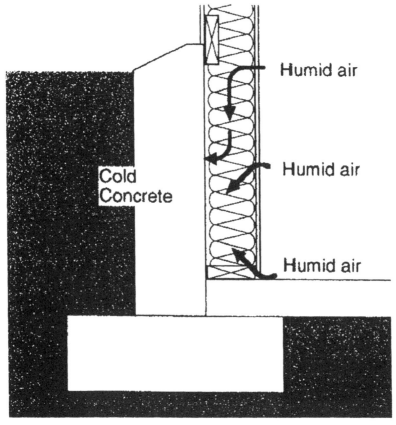

FIGURE 84. A detail of the "garden level" units showing a wood frame wall constructed to the interior of the concrete foundation assembly.

2.3 Proposed Corrective Measures

Following recommendations should alleviate both the problems induced by air pressure differences and by the thermal bridging at "garden level" units:

- seal the air distribution system particularly the boot connections with mastic and fiber reinforcement (Figures 85 and 86)
- seal the plenum returns of air handlers installed in dropped ceilings
- balance the air distribution system by providing transfer grills (Figure 87) reduce the operation of exhaust fans such as bathroom fans via the use of time delays linked to lights
- reduce the thermal bridging following measures shown in Figures 88 and 89

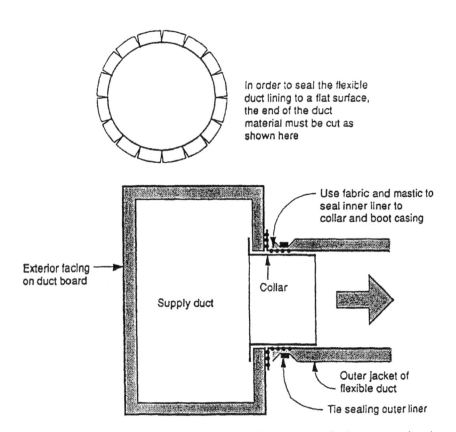

FIGURE 85. Remedial measures for the air distribution system, the boot connections in particular.

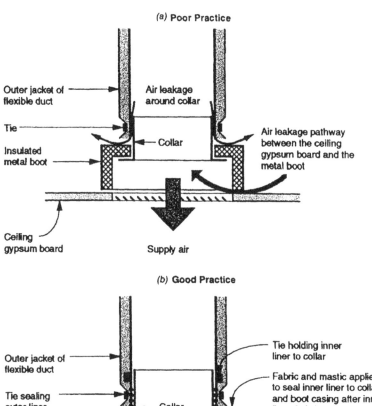

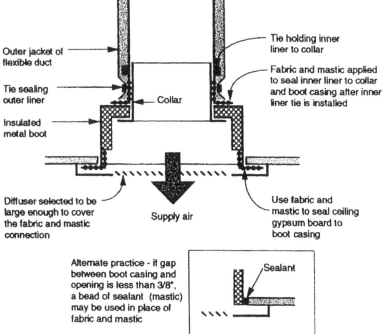

FIGURE 86. The boot connections in the air distribution system: (a) bad, (b) good practice.

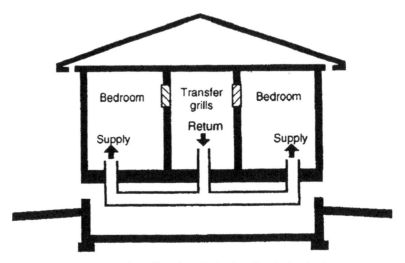

FIGURE 87. Transfer grills assist in balancing the air distribution system.

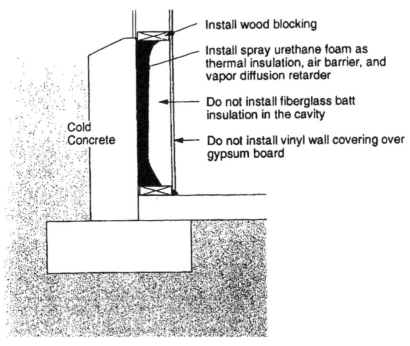

FIGURE 88. Vertical section through wood frame wall constructed to the interior of the concrete foundation assembly in the "garden level" units.

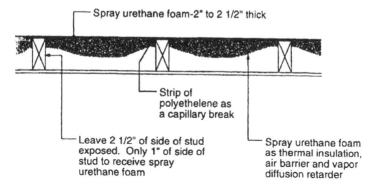

FIGURE 89. Horizontal section through wood frame wall constructed to the interior of the concrete foundation assembly in the "garden level" units.

3. AIR LEAKAGE CONTROL FOR RETROFITS FOR HIGHRISES[65]

Two buildings were selected for the demonstration of air leakage control. Tests were conducted to verify and characterize these buildings before and after the air-sealing work:

- whole building airtightness tests
- indoor air quality
- monitoring energy and power consumption

The following summarizes the field data and analysis. Building A is a well maintained and fully occupied 21-story, 240-suite apartment building located in Ottawa on open and flat terrain. Building B is a 10-story, 95-suite apartment building located in a suburb of Toronto.

The following air sealing measures were performed in building A:

- caulking and weatherstripping of windows, entry doors, and balcony doors
- sealing of floor/wall, roof/wall, and wall corner joints, exhaust fan housing from interior, canopy joints, electrical receptacles
- sealing of garbage chutes, corridor stairway doors, and elevator shafts

The cost of these repairs was $41,500. Simple payback period using the first year savings was 4.3 years. Inspection after two years showed the energy savings about 12 percent per year. Sealant and weather stripping were in good condition and there were no more complaints about cold drafts.

---

[65]Based on Woods, Pakesh and Sinnige article in JTIBE, Vol. 19, Oct. 1995.

The following air-sealing measures were performed in building B:

- caulking and weather stripping of windows, entry doors, and balcony doors
- sealing of floor/wall, roof/wall, and wall corner joints, exhaust fan housing from interior, canopy joints, electrical receptacles
- sealing of garbage chutes, corridor stairway doors and elevator shafts
- isolation of the ground floor common rooms

The cost of these repairs was $30,000. Simple payback period using the first year savings was 4.2 years. Inspection after two years showed that sealant and weather stripping were in very good condition.

The airtightness of the building envelope was improved 32 percent to 38 percent. The peak heating space demands were significantly reduced and there were no negative effects of the air-sealing on the IAQ or comfort in both buildings.

4. SPF USED AS THE AIR-SEALANT IN DIFFERENT APPLICATIONS[66]

In contrast to the previous case studies, this section provides a short overview of different applications of the BSF type of SPF.

4.1 Air Leakage at Steel Deck-Wall Intersection

As a part of a performance contract in twenty-one schools of the Muskoka Board of Education, eighteen schools had two-component froth foam[67] used to seal the roof-wall intersection (Figure 90). To this end, deck flutes were punctured, and the foam was injected. As the plane of the air barrier is not the steel deck but the roofing membrane, the deck can be punctured without any detriment to the air barrier function of the roofing system.

This concept is also illustrated in a re-designed detail of a new building. Under a contract for the troubleshooting of a new design, Sartor (1995) introduced several modifications that are shown in Figure 91. SPF was used both in the flutes of the steel deck and around the metal track placed on the steel studs. This design of the detail was selected for reproduction because it uses a rubberized asphalt membrane (called WAM membrane in this monograph) and BSF in combination with the sealing compound to ensure airtightness of the joint.

---

[66]This section is based on information and photographs provided by Mr. A. Woods, president of CAN-AM Building Envelope Specialists, Mississauga, Ontario (see acknowledgements).
[67]One 100 lb. froth kit was used for approximately 600 ft. of roof/wall intersection. This case study is described in Flexible Products Manual.

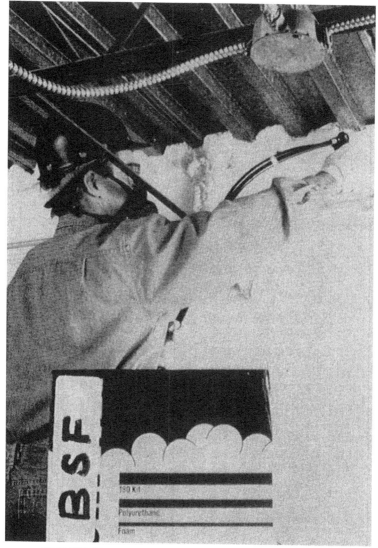

FIGURE 90. Application of BSF in the wall-roof intersection.

Sartor (1995) explains the need for the WAM membrane as follows:

We added "peel and stick" the rubberized asphalt membrane flashing over the wood coping, bonded to the roofing membrane, to protect the top of the concrete block structure from premature deterioration, and to avoid possible water penetration to interior. This consideration is often overlooked in design of

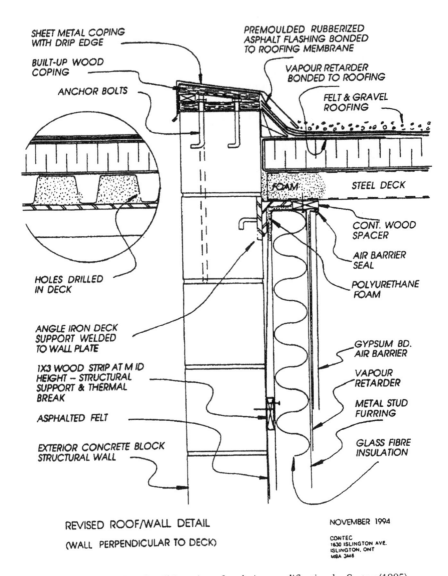

SHEET METAL COPING
WITH DRIP EDGE

BUILT-UP WOOD
COPING

ANCHOR BOLTS

PREMOULDED RUBBERIZED
ASPHALT FLASHING BONDED
TO ROOFING MEMBRANE

VAPOUR RETARDER
BONDED TO ROOFING

FELT & GRAVEL
ROOFING

HOLES DRILLED
IN DECK

ANGLE IRON DECK
SUPPORT WELDED
TO WALL PLATE

1X3 WOOD STRIP AT MID
HEIGHT — STRUCTURAL
SUPPORT & THERMAL
BREAK

ASPHALTED FELT

EXTERIOR CONCRETE BLOCK
STRUCTURAL WALL

FOAM   STEEL DECK

CONT. WOOD
SPACER

AIR BARRIER
SEAL

POLYURETHANE
FOAM

GYPSUM BD.
AIR BARRIER

VAPOUR
RETARDER

METAL STUD
FURRING

GLASS FIBRE
INSULATION

REVISED ROOF/WALL DETAIL

(WALL PERPENDICULAR TO DECK)

NOVEMBER 1994

CONTEC
1630 ISLINGTON AVE.
ISLINGTON, ONT
M9A 3M8

FIGURE 91. Roof-wall junction after design modification by Sartor (1995).

copings, the designer usually depending on the sheet metal flashing to act as a membrane. Of course it will not, and cannot, be a *leakproof* membrane, since the sheet metal joints will always leak, even when caulking with an appropriate sealant material. This is because the joints move considerably, even daily, with temperature changes.

Sartor explains that plywood is preferred over dimensional lumber because of lower warpage and shrinkage potentials, and that long anchor bolts are used to prevent the top block course from becoming dislodged during construction. These bolts are 9 mm in diameter (3/8 in), spaced 400 mm (16 in), and staggered. Sartor (1995) also ensures that in the modified design there is no thermal bridging between block and metal stud furring. A wood strip (25 mm × 75 mm nominal dimensions) is placed in the mid-height and anchored on both sides, to the wall and to the studs. The need for the wood strip is explained as follows:

> This transfers the wind load from the gypsum board air barrier to the block wall structure, through the metal studs, keeping the gypsum board furring from distorting under severe wind loads and gusting.

Sartor (1995) notes:

> Since the edge of the deck is welded to the structural steel anchored to the concrete block, it will reflect outside temperatures. Therefore, the top track is insulated from the steel roof deck with a wood strip to prevent thermal bridging here.

A caulking bead is applied to the wood strip to provide continuity of the air barrier function. Since the SPF acts as the air seal, this caulking bead is provided as the second line of defense. If the caulking was to perform air-seal function, we would require a proper joint design (dimensions and support for the caulking bead) or select an elastic gasket. As the second line of defense, however, one may use acoustic caulking.

In the direction parallel to the roof deck, pieces of wood strapping are long enough to be anchored to the second flute. SPF is used to fill the space between the wood strapping and the space in the flutes above the wall.

4.2 Party Wall and Top Plate Sealing in Attics

Figure 92 shows the roof of raw housing and the typical pattern of melting snow caused by heat transferred along the concrete block wall.

Both one- and two-component BSF types of SPF are used to seal the path of the air transfer along the party walls. Block cores are drilled or punctured with a special hammer and the foam is applied to seal the block cores as well as the space between the block and the drywall. As shown in Figure 93, BSF may be applied along the joist adjacent to the party wall.

FIGURE 92. Melting snow marks areas with increased heat transfer.

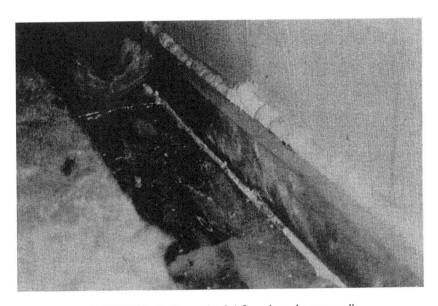

FIGURE 93. Sealing path of airflow along the party wall.

Older homes (pre-1970) in Canada did not have the continuity of the air-vapor barrier specified for the top plate. As wet wood was frequently used, it often shrank allowing air passage to attics along internal and external walls. BSF is also used to seal the air path in these locations.

4.3 Roof-Wall Intersection

While ventilation of attic space is needed for temperature equalization and periodic or seasonal removal of moisture (e.g., that accumulated during the winter) in the attic space, excessive ventilation may be unwanted. It may lead to ingress of snow, as was the case in the Scarbrough School Board. Initially, the weatherization involved four schools, but with fuel savings of 35 percent, 17 percent, 9.5 percent and 5.6 percent respectively, recorded in the first year after air-sealing the attics, fifty-six other schools were added to this project.

4.4 Condensation Control

A building, air conditioned at 50 percent RH and operating under positive pressure because of many computer rooms, showed signs of excessive condensation. Examination showed that foil-faced board insulation applied in precast cladding panels had gaps. Two-component froth BSF was used to seal joints between the insulation sheets as well as to fill the window mullions. Because of difficult access, tubing was used to connect the foaming gun and the space where the foam was to be injected (Figure 94).

4.5 Improving Indoor Environment by Reducing Stack Effect

A 28-story building in Toronto was plagued by the percolation of cooking odors upside the walls. The solution, approved by the local building authority, was to remove the rubber baseboard and seal gaps around precast anchors. Two-component, froth BSF was used for air sealing.

In another 28-story building, stack effect was drawing dirt from three levels of underground parking up to the sixteenth floor, through the elevator shaft. The carpet edges were quite dirty, so the baseboards were foamed in all hallways.

To reduce the air-flow further, the infill masonry walls between concrete sheer walls in parking areas were also foamed. (The mineral fiber originally placed in these joints was dirty, indicating presence of air flow.) High expansion,[68] one-component BSF was applied in these joints.

---

[68]Approximately 200 × 33 oz. kits were used.

FIGURE 94. Injecting BSF into window mullions.

4.6 Improving Wall Durability by Reducing Stack Effect

Two 15-story apartment buildings showed signs of brick deterioration characteristic of an exfiltration of moist indoor air (Figure 95).

Primarily, two-component froth BSF was used for air sealing. Using a 55,000 c.f.m. blower door to test air leakage rate before and after the BSF application, a 35 percent reduction of air leakage was measured.

In a similar case of condensation on eight upper floors of the thirteen story building in Toronto, two-component SPF was injected[69] through drywall to seal the covers of the precast columns.

4.7 Duct and Rim Joist Sealing

Use of BSF to seal duct penetration is rapidly becoming popular because of speed and convenience of workmanship. When a small quantity of foam is applied in the form of a bead or strip, the smoke development index is normally less than fifty. Similarly, when a small quantity of BSF is used for strips along the joist line, the standard requirements of fire protection are usually met. When the whole area of the rim joist is covered, a thermal barrier will normally be required by a local Fire Marshall.

---

[69]Approximately 5,000 lb. were used in this application.

4.8 Rehabilitation of Windows

Figure 96 shows detail of new wood window being inserted in the existing window frame. One-component BSF is used on both sides of the existing wood frame, to ensure continuity of air-sealing within the existing wall as well as with the new window frame. The weather protection is achieved by a piece of dimensional lumber attached to the existing wood frame, which also secures positioning of the inserted window. Normally, a bead of silicone caulking would be applied to the back surface of this lumber and the new frame inserted tight, to expel excess caulking. Then, the cavity between the old and new frame is filled with BSF (Figure 97), normally in a few passes (a minimum of two).

4.9 Other Applications: Soil Stabilization, Water Flashings, Exhaust Boxes, Passage of Insects

In one case of winter construction involving piled foundation, a two-component BSF foam was applied on the pile and soil surfaces to prevent major changes in soil moisture content.

FIGURE 95. Two, 15-story, apartment buildings showed signs of brick deterioration.

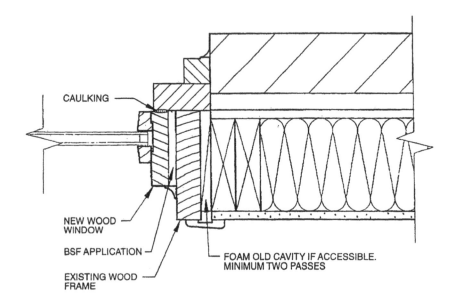

FIGURE 96. Rehabilitation of an old wood window.

FIGURE 97. Application of BSF to fill the space between two window frames.

The same BSF was applied to form a second line of defense flashing. This flashing was placed between vertical aluminum curtain wall and granite siding (Figure 98). The number of projections and difficulty in achieving adhesion in the cold weather made use of a rubber flashing improbable. The BSF was then covered with elastomeric roofing coating material to provide a long-term durability.

FIGURE 98. Sealing penetration with a sloped foam, covered with elastomeric coating to function as a "flashing."

Finally, another case when insects (cluster flies) were entering a geriatric center through a hidden passage in the external wall was also alleviated with the use of one component BSF. The flies were entering weep holes in the brick veneer, following the air cavity to a window sill and entering through penetrations for incremental heating units. During this air-sealing project, many other gaps or cracks were also filled. Payback was obtained through energy saving within three years.

5. CONCLUSIONS FROM THESE CASE STUDIES

Many North American houses are not designed. They are built. While building, we follow tradition more than science, and we try minute alterations. We modify materials, workmanship, construction details, or other seemingly unimportant elements of the construction process. Then, as in the above discussed examples, the problems caused by these minute changes make us believe that these details were important. This situation happens each time, when we analyze only the detail itself and forget interactions between different elements of the system. In other words, we fail when we lose track of the holistic approach.

Many case studies reported in this chapter deal with moisture-originated problems. For instance, a poor design of the environmental control of the building envelope in a computer center facility resulted in a cladding failure. Evidently, the requirements for pressurized and humidified interior space call for specialized knowledge. In another case, a performance failure was caused by disregarding interactions between the HVAC and the building envelope. In both these cases, the condensation of moisture happened either at a place or in a manner not readily evident to the construction professionals. The architect or designer usually considers a cross-section of the wall under typical outdoor and indoor service conditions. The redistribution of air in the indoor space and the interaction between the HVAC and the building envelope are rarely considered.

Several case studies were focused on air leakage control through the building envelope. An overview of various BSF type foam applications showed the benefit of their use. The reason for selecting these case studies for the closing chapter of this monograph is simple—the proof of pudding is in eating. The versatile nature of the foamed in place products is shown in the context of different demands that are placed on the repair and rehabilitation of these buildings.

Recommended Procedure for Determination of Long-Term Thermal Resistance of Closed-Cell Thermal Insulating Foams[70]

1. SCOPE

THIS PROCEDURE DEFINES the long-term thermal resistance (LTTR) of a foam product as the value measured after 5-year storage in a laboratory environment and provides means for its prediction based on an accelerated laboratory test.

This procedure, based on ASTM standard test method C1303, can be applied to a wide range of preformed as well as field applied insulating foams. This procedure estimates the change in the thermal resistivity of insulating foam products by means of slicing and scaling.

This procedure is applicable to products with either permeable or semi-permeable facers. The user of this procedure shall determine if the low permeance of the surface layer or the facer limits the use of this procedure for the specific application.

This procedure specifies reference time, sampling and testing requirements for the application of ASTM standard test method C1303 to determine LTTR for closed-cell foams such as extruded polystyrene, sprayed polyurethane, polyisocyanurate or other foams addressed by ULC thermal insulation standards.

This procedure does not purport to address all the safety concerns, if any, associated with its use. It is the responsibility of the user to establish appropriate safety and health practices, and determine the applicability of regulatory limitations prior to use.

---

[70]This is a draft proposal developed by a Task Group (see Appendix) for the Thermal Insulation Committee at the Underwriters Laboratory, Canada (a national standard writing organization in area of thermal insulating systems).

255

2. TERMINOLOGY

2.1 Definitions

For definitions of terms and symbols used in this procedure and C1303 standard, refer to ASTM standard C168: Standard Terminology Relating to Thermal Insulating Materials.

2.2 Description of Terms Specific to This Procedure

Long-term thermal resistance (LTTR)—A design thermal resistance of the product, measured at standard laboratory conditions (22 $\pm$ 2°C and 45 $\pm$ 5% RH) equivalent to 5-year storage in a room (24 $\pm$ 4°C and 45 $\pm$ 20 percent RH).

Aging factor—A ratio between the average thermal resistivity at the end of the prescribed aging period to the average initial thermal resistivity.

Scaling equation—For any foam product with identical gas diffusion characteristics, two different specimens with thickness L_1 and L_2 will reach the same degree of aging in time t_1 and t_2 such that $t_1/t_2 = (L_1/L_2)^2$. This relationship is defined as the scaling equation.

Scaled aging time—The time at which specimens of different thickness cut from the same foam sample are in the same state of aging as that of the foam sample.

Testing points—A time estimate for a thin slice when a thermal test performed on the slice will correspond to the same degree of aging of the full board with specified thickness and 5-year reference period.

3. SIGNIFICANCE AND USE

This procedure includes a check to determine whether the effect of low permeability of the surface layer is within specified limits.

This procedure allows testing a 5-year-old product to replace the estimated LTTR with the measured value.

Where thermal insulating products are manufactured under field conditions, the nominal thicknesses for which LTTR is determined shall be 25, 50 and 75 mm. The middle thickness, 50 $\pm$ 5 mm, shall be used for testing.

Where preformed thermal insulating products are tested LTTR shall be determined for the actual thickness of the supplied product, and for 25-mm thicker and 25-mm thinner, if applicable. For products manufactured with a thickness range, the middle thickness shall be selected for testing.

4. SAMPLING

Select a minimum of three samples, at least two hours apart, from the production run (or the field manufacture). Deliver these samples to the laboratory such that the lab receives them in 10 days or less from the date of manufacture. The actual size of the sample shall be agreed between the testing laboratory and the proponent. However, the minimum dimension of each sample shall be 600 mm × 1200 mm (2′ × 4′) for preformed products and 900 mm × 900 mm (3′ × ′3′) for a field manufactured product (e.g., sprayed polyurethane foam).

For a field-manufactured product, select a 100-mm (4″) thick slab sprayed on a rigid substrate to avoid deformation of the test sample (due to warping). A 12-mm (1/2″) thick plywood board with 20 mm–30 mm (1″–1.5″) spray polyurethane foam backing on the other side has been found to provide satisfactory test samples.

5. THERMAL TESTING

5.1 Sequence of Testing for a Laboratory Equipped with 600 mm × 600 mm (24″ × 24″) as Well as 300 mm × 300 mm (12″ × 12″) Heat Flow Meter (HFM) Apparatus

After delivery of the samples to the testing laboratory, cut three specimens, 600 mm × 600 mm (24″ × 24″), one from each board. If the major surfaces of the test specimens are not parallel, a maximum thickness of 5 mm (1/5″) shall be removed from both surfaces to make them parallel. In no case the total thickness so removed shall exceed 15% of the product thickness.

The initial thermal resistance and resistivity measurements on the above three specimens shall be completed within 7 to 14 days after the production date. An average of the three specimens shall be calculated as the initial thermal resistance of the sample given for testing. Calculate the initial thermal resistivity as an average of these three specimens.

Denote the day on which the first of the three specimens is tested as day 1, for subsequent calculation of the testing points. On the same day, cut that 600 mm × 600 mm specimen into four 300 mm × 300 mm specimens. From each of two of the smaller specimens, slice two layers adjacent to the surface (called "surface layers"), with thickness between 6 mm and 12 mm and two layers with the same thickness as the surface layers from the middle part (called "core layers"). There shall be a total of 8 thin layers, four surface layers and four core layers, available for subsequent testing. If the required number of core layers cannot be sliced from two smaller specimens (as restricted by the total thickness of the 600 mm × 600 mm specimen), the remaining two smaller specimens shall be used for completing the total 8 thin layers for subsequent testing.

5.2 Sequence of Testing for a Laboratory Equipped with Only 300 mm × 300 mm (12″ × 12″) HFM Apparatus

After delivery of the samples to the testing laboratory, cut five specimens, 300 mm × 300 mm (12″ × 12″), one from each board. If the major surfaces of the test specimens are not parallel, a maximum thickness of 5 mm (1/5″) shall be removed from both surfaces to make them parallel. To comply with ASTM Standard C518, the thickness of these test specimens shall not exceed 38 mm (1.5″).

The initial thermal resistivity measurements on the above five specimens shall be completed within 7 to 14 days after the production date. An average of the five specimens shall be calculated as the initial thermal resistivity of the samples given for testing. Calculate the initial thermal resistivity as an average of these five specimens.

On the same day, denoted as Day 1 for subsequent calculation of the testing points, select two of the five specimens and from each of these, slice two layers adjacent to the surfaces (called "surface layers") with thickness between 6 and 10 mm, and two layers with the same thickness as the surface layers from the middle part (called "core layers"). Thus there will be four surface layers and four core layers available for further testing. If the required number of core layers cannot be sliced from two specimens (as restricted by the total thickness of the 300 mm × 300 mm specimens), a third specimen shall be used for completing the total 8 thin layers for subsequent testing.

5.3 Testing the Thin Layers

Measure and record the thicknesses of each of the eight thin layers. If the thin layers have thicknesses within 5 percent of the mean value of all the thin layers, the four surface layers and the four core layers may be tested as a stack each. The surface layers and the core layers shall not be mixed to form a stack. If the thicknesses of the layers are not within 5 percent of their mean value, they must be tested separately.

Measure the initial thermal resistivities of the layers. It is essential that this be done within 2 hours of cutting because the thin layers will age very rapidly.

From the thickness of the layers (which can be the average thickness if they are stacked), calculate the testing points when the layers must be re-tested to correspond to the five-year aging of the product at the various thicknesses.

The scaling equation shall be used for this purpose. As an example, for an insulating product manufactured in the field and for a 10 mm thin layer the testing points for various product thicknesses shall be as follows.

Product Thickness, mm	Testing Point, Day
25	292.0
50	73.0
75	32.4

If requested by the proponent, two additional core slices with the thickness in the range of 25 mm to 38 mm may also be included in the testing program. *(Note: It is customary to select a 25 mm to 38 mm thick core layer for detailed testing of the first stage of aging.)*

6. CALCULATING LTTR

Within 24 hours of the prescribed testing points, i.e., the calculated scaled aging time, remeasure the thermal resistivity of each surface layer and of each core layer or the stacks of surface and core layers. Calculate the average aging factors of the surface layers and of the core layers. If the measurements are done using a stack of four, the aging factor will be derived from the thermal resistivity of the whole stack. (The aging factor is derived by dividing the thermal resistivity of the layers at the testing point by the initial thermal resistivity, as obtained from the second paragraph of Section 5.2.)

If the difference between the averaging aging factor of the surface layers and that of core layers does not exceed 12 percent of their mean value, use the higher of these two values as the effective aging factor. (See Limitations for Use Scaling Equation in the Appendix.) If the difference is more than 12 percent, the test is considered invalid, and cannot be used for determining LTTR.

Establish LTTR as a product of the average initial thermal resistivity of the product (see the second paragraph of Section 5.1 or the second paragraph of Section 5.2), the thickness under consideration and the effective aging factor as defined in Section 6.2.

7. REPORTING

The test report shall include the following information, including references to applicable tests methods:

- the date of the report
- the name, address and identification of the testing laboratory

- the manufacturer of the material, the date of manufacture and the date of receiving samples
- number of samples received and the number of specimens tested in respective categories
- the name or identification of the material tested and description of facers (if any)
- the method of specimen preparation and the aging conditions
- the type and size of the thermal test apparatus and the method of its calibration
- the mean test temperature, the temperature difference and the age of each specimen at each test time
- the time when the initial thermal resistivity of the full thickness product was measured
- the thickness of surface/core slices and whether they were tested individually or stacked
- the average thermal resistivity of the surface and core layers

8. APPENDIX: RATIONALE FOR KEY ELEMENTS IN THIS RECOMMENDED PROCEDURE

8.1 Background

The thermal resistance of thermally insulating foam products may change during their service life, mainly because of the changes in the composition of the gas contained within the cells of the foam, due to diffusion. This phenomenon is referred to as aging. Since aging is a very slow process and occurs over many years of service life, from a practical point of view, the aging process must be accelerated to evaluate the long-term thermal resistance (LTTR) of the insulating materials. This can be done either by means of elevated temperature or by the use of thin layers of the product. The former approach changes the permeability and solubility coefficients of different gases to such a degree that the rate of aging measured under elevated temperature may not be representative of the aging under service conditions. In the latter approach, this is not an issue.

The technique, known as "Slicing and Scaling" relies on the fact that the rate of gas diffusion is inversely proportional to the square of the thickness of the product. If the thickness is halved, the aging is four times as fast. Thus, a 10-mm thick slice will be in the same state of aging after 73 days as a 50-mm thick board will be after 5 years.

In 1987 a Canadian expert group published LTTR values for different closed cell foams (Kabayama, 1987). Nevertheless, changes of the blowing agents (Kumaran and Bomberg, 1990) required a review of these values as well as development of a uniform approach.

The CPIA/NRC project (Bomberg and Kumaran, 1994) verified that the thermal transmission properties of insulating foam products aged in the field for 2-1/2 years were not significantly different from those for the same products aged in the laboratory. From this and similar information it was concluded that LTTR can be defined just based on the aging carried out under standard laboratory conditions.

8.2 Definition of LTTR

The long-term thermal resistance (LTTR) of a thermally insulating foam product is defined as its thermal resistance measured under standard laboratory conditions (22 ± 2°C and 45 ± 5 percent RH) after 5 year storage in a room (24 ± 4°C and 45 ± 20 percent RH). The LTTR is a design property, introduced for comparing different foam products to one another.

The consensus on the design thermal resistance was established in two steps. First, the Canadian Plastics Industry Association and the manufacturers agreed that LTTR shall be defined as the time weighted average of thermal resistance over 15 years at a given thickness. Secondly, it has been demonstrated that the average resistance over a given period of time is equal to the value measured at a reference time obtained by dividing the specified period by a number ≈3. Thus, the selected reference time is 15/3 = 5 years.

8.3 Testing Methodology

The recommended procedure for the determination of LTTR of closed cell plastic foams is based on ASTM standard C1303 that measures aging of thin layers (slices). However, in few cases does any slice cut from the foam truly represent the full foam. Density and cell structure may vary across the thickness of the foam and densified skin or facers may also affect the rate of aging.

Since the uncertainty of thermal characteristics determined on small pieces of the foam is larger than the variability seen among standard laboratory samples, the methodology is divided into three steps:

1. Determination of the mean initial thermal resistance of the product
2. Determination of the aging factor as the ratio between the thermal resistivity at the specified time of aging to its initial value
3. Calculating LTTR as the product of the initial thermal resistance and the aging factor

8.4 Limitations for the Use of a Scaling Equation

The traditional approach to slicing method is to cut a thin layer from the middle of the foam (core) and test its aging as a function of time. However, the density gradient across the thickness or modified cellular structure obtained during lamination process may retard the rate of aging of the foam. Hence, the proposed procedure prescribes the measurements on both cure and surface layers. However, this has some implications.

As the surface layers are sliced, a new surface that facilitates aging is created—the one towards the core. In the original product this cut surface was not participating in the aging process. Hence, the accelerated aging on the surface layers may not be truly representative of the product. The implications of this may be examined as follows.

Let the measured aging factors according to the proposed procedure for the surface and core layers be a_s and a_c, respectively, and let $a_s > a_c$. If there were only one of the surfaces of the surface layer participating in the aging process, its effective aging factor A_s may be approximated as:

$$A_s = 2(a_s - a_c) + a_c \qquad (1)$$

The effective aging factor, a_{eff}, for the product may be approximated as:

$$a_{eff} = (A_s + a_c)/2 \qquad (2)$$

Substitution of Equation (2) into Equation (1) gives, $a_{eff} = a_s$. That means the surface layer controls the aging of the product.

The reverse situation cannot be ruled out where $a_s < a_c$. Then it can be shown through the same argument that $a_{eff} = a_c$.

In reality the effective aging factor of the product must be a combination of both aging factors, a_s and a_c. This combination depends on the nature of the foam product. As a conservative estimate, the present procedure recommends adopting the higher of the two as the effective aging factor. However, if the difference between the two is more than 12%, this conservative estimate should be reexamined.

Once cut, the thin layers age very rapidly. The aging factor, derived as the ratio of thermal resistivity at the testing points to the initial resistivity, can only be applied to the full board, if the initial thermal resistivity of the board and slices corresponds to the same scaled age, measured from the time of cutting. Say 10 mm layers were cut from a 50 mm product, and tested 8 hours later. Applying the scaling equation, we see that a retained 50 mm board will be at the same scaled age as the 8 hour slices after $(50/10)^2 \times 8$, or 200 hours (8.3 days). Eight days becomes the time at which the initial thermal resistivity of the full thickness product should be measured.

ACKNOWLEDGEMENTS

This procedure was written by a Task Group consisting of Dr. M. K. Kumaran, IRC/NRC, chairman; Martin Hofton, Owens Corning Inc.; Dr. Michel Drouin, Exceltherm Inc.; Gary Chu, Dow North American Inc.; Andre St. Michel, BASF; Ron Waters, CCMC/NRC and Dr. Mark Bomberg, ex-officio, as chairman of the ULC standardization committee.

Glossary of Terms and Abbreviations Used by the SPF Industry[71]

A–Side (A–Component) One component of a two component system; for polyurethane foam and coatings the isocyanate component.

Absolute Humidity The actual concentration of water vapor in air. May be expressed in units of mass, grams of moisture per kilogram (pound) of dry air, or as a partial pressure in Pa or inches of mercury (in Hg).

Accelerator A chemical additive to coating or polyurethane foam systems used in relatively small amounts to increase the speed of the reaction or to decrease the time required to cure or dry.

Acrylic Coating A coating system based on an acrylic resin; generally, a "water based" coating system which cures by coalescence and drying.

Acrylics Resins resulting from the polymerization of derivatives of acrylic acids, including esters of acrylic acid, methacrylic acid, acrylonitrile, and their copolymers. They can be carried in a water or solvent solution and they are film forming materials.

Adhesion The degree of attachment or bonding of one substance to another. (Though interlaminate adhesion is used, COHESION is a more appropriate term to describe the bonding between different applications of the same substance.)

Aggregate Any mineral surfacing material (crushed stone, gravel, washed gravel, roofing granules, etc.).

Aging An effect of time on materials that are exposed to an environment for an interval of time.

Aliphatic (Urethane) An organic polymer containing straight or branched chain arrangements of carbon atoms. Coatings usually have better gloss, color retention, and weathering ability.

Alligatoring Pattern cracking of a coating or mastic resembling the pattern of an alligator skin (See CHECKING).

---

[71]Based on SPFD terminology document.

265

Anodic When two metals are connected in an electrolyte, they will form a galvanic cell, with the higher metal in the galvanic series being the anode.

ANSI American National Standards Institute.

APA American Plywood Association.

Application Rate The quantity (mass, volume, or thickness) of material applied per unit area.

Apron Flashing A term used for a flashing located at the juncture of the top of the sloped roof and a vertical wall or steeper-sloped roof.

Architectural Panel A metal roof panel, typically a double standing seam or batten seam; usually requires solid decking underneath and relies on slope to shed water.

Area Divider A raised, flashed assembly (typically a single or double wood member attached to a wood base plate) that is anchored to the roof deck. It is used to relieve thermal stresses in a roof system where an expansion joint is not required, or to separate large roof areas (sometimes between expansion joints). It may also be used to facilitate installation of tapered insulation.

ARMA Asphalt Roofing Manufacturers Association.

Aromatic (Urethane) An organic polymer usually containing one or more benzene rings structures. Aromatic urethane coatings are usually tougher and cost less than aliphatic urethanes.

Aromatic Solvents Hydrocarbon solvents comprised of organic compounds which contain an unsaturated ring of carbon atoms, including benzene, xylene, toluene and their derivatives.

ASA American Subcontractors Association.

ASC Associated Specialty Contractors.

ASHI American Society of Home Inspectors.

ASHRAE American Society of Heating, Refrigerating and Air-Conditioning Engineers, Inc.

Asphalt A dark brown or black substance found in a natural state or, more commonly, left as a residue after evaporating or otherwise processing crude oil and petroleum. It consists mainly of hydrocarbons. Asphalt products are available for hot or cold application. Asphalt will dissolve in mineral spirits.

Asphalt Emulsion A mixture of asphalt particles and an emulsifying agent such as bentonite clay and water. These components are combined by using a chemical or a clay emulsifying agent and mixing or blending machinery.

Asphalt Felt An asphalt saturated and/or an asphalt coated felt (See FELT).

Asphalt Roof Cement A trowelable mixture of solvent based bitumen, mineral stabilizers, and other fibers and/or fillers. Classified by ASTM Standard D 2822-91 Asphalt Roof Cement, and D 4586-92 Asphalt Roof Cement, Asbestos Free, Types I and II.

ASTM American Society for Testing and Materials.

B-side (B-component) One component of a two component system; for polyurethane foam and coatings the isocyanate component.

Back Rolling Rolling wet coating behind a spray or roller application, to insure better coverage on rough surfaces.

Base Coat The first coat of a multi-coat system. This should be applied the same day as the SPF.

Bird Bath Random, inconsequential amounts of residual water on a roof membrane.

Bitumen A class of amorphous, dark brown to black (solid, semi-solid, or viscous), high molecular weight hydrocarbons derived from petroleum refining (asphalt) or coal reduction (coal tar).

Bleeding (1) The diffusion of coloring matter through a coating from its substrate (such as bleeding of asphalt mastic through coating). (2) The absorption of oil or vehicle from a compound into an adjacent porous surface.

Blister An uplifting of coating or polyurethane foam caused by an enclosed pocket of gas entrapped between coating passes, foam and coating, foam and substrate, or within the foam itself. Caused by the DELAMINATION (cf. also) of one or two components in an insulation or roofing system.

Blowing Agent A gas (or a substance capable of producing a gas) used in making foamed materials (see CAPTIVE BLOWING AGENT).

Blow Holes Holes in the coating and/or polyurethane foam surface caused by escaping gas or blowing agent that was produced during the foaming application.

BOCA Building Officials and Code Administrators, International. One of three model building codes in the U.S.

Bond, Chemical Adhesion between surfaces, resulting from a chemical reaction or cross-linking of polymer chains.

Bond, Mechanical Adhesion between surfaces resulting from interfacial forces or a physical interlocking.

Building Code A system of principles or rules governing the design and construction of buildings. Local governments generally adopt or modify one of the model building codes. (See: BOCA, ICBO, SBCCI, IBC, ICABO in the U.S. or NBC in Canada.)

Building Envelope   The exterior shell of a building designed for environmental protection and structural requirements (includes roofs, walls, basement walls, slab on ground, etc.).

Built-Up Roof (BUR)   A roofing membrane consisting of alternating applications of bituminous impregnated felts or fabrics and hot/cold mopped bitumen. The membrane is generally surfaced with aggregate.

Butyl Coating   An elastomeric coating system derived from polymerized isobutylene. Butyl coatings are characterized by low water vapor permeability.

CABO   Council of American Building Officials. A code agency.

Calorimeter   An apparatus for measuring quantities of heat developed by combustion.

Cant   A beveling of polyurethane foam at a right angle joint for strength and water run-off.

Cant Strip   A beveled strip used under flashings to modify the angle at the point where the roofing or waterproofing membrane meets any vertical element.

Cap Flashing   Usually composed of metal, used to cover or shield the upper edges of the membrane base flashing, wall flashing or primary flashing.

Capacitance Meter   A device used to detect moisture or wet materials by measuring the ratio of the change to the potential difference between two conducting elements separated by a nonconductor.

Capillary Action, Capillarity   Movement of liquid in the interstices of insulation or other porous material as a result of surface tension.

Catalyst   An ingredient in a coating or polyurethane foam system, which initiates a chemical reaction or increases the rate of a chemical reaction.

Captive Blowing Agent   An insulating gas, normally better insulator than air, that is used during manufacturing but retained by the foamed product improves its thermal resistance (e.g., HCFC, HFC).

Caulk   A flexible waterproofing material used to seal cracks, seams, or small breaks in a waterproofing or an air seal system. Usually supplied in tubes and applied with a caulking gun (See SEALANT).

Cavitation   The vaporization of a liquid under the suction force of a pump. Usually due to inadequate flow to a pump; the vaporization can create voids within the pump or the pump supply line. In polyurethane foam spray pumps, cavitation will result in OFF-RATIO FOAM (see also).

Cavity Wall An exterior wall, usually of masonry, consisting of an outer and inner width separated by a continuous air space (see also FRAME WALL).

Cellular Describes a composition of plastic or rubber with relative density decreased by the presence of cells dispersed throughout its mass. In closed-cell materials, the cells are predominately separate from each other. In open-cell materials, the cells are predominately interconnected.

Centipoise (CPS) A unit of measure of absolute viscosity. (Note: The viscosity of water is one centipoise. The lower the number, the less the viscosity.)

Chalking The formation on a surface of a powdery substance due to weathering.

Checking A defect in a coated surface characterized by the appearance of fine fissures in all directions. Designated as "surface checking" if superficial; or "through checking" if extending deeply into the coating or to an adjoining surface.

Chemical Resistance The ability to withstand contact with specified chemicals without a significant change in properties.

Chlorinated Rubber Resin formed by the reaction of rubber with chlorine to form a coating (i.e., primer or hypalon) or single ply membrane. Unlike rubber, the resulting product is readily soluble and yields solutions of low viscosity. Commercial products generally contain about 65% chlorine. It has good chemical resistance properties. It tends to cobweb when sprayed.

Coal Tar A dark brown to black hydrocarbon obtained from the destructive distillation of coal. Used in built-up roofs or in below grade construction as a primary waterproofing agent.

Coalescence The formation of a film of resinous or polymeric material when water evaporates from an emulsion or latex system, permitting contact and fusion of adjacent latex particles. Action of the joining of particles into a film when volatile compounds evaporate.

Coarse Orange Peel Surface Texture A surface showing a texture where nodules and valleys are approximately the same size and shape. This surface is acceptable for receiving a protective coating because of the roundness of the nodules and valleys.

Coating A layer of material spread over a surface for protection and decoration. Coatings for polyurethane foam are liquids, semi-liquids, or mastics; spray, roller, or brush applied; and are ELASTOMERIC.

Cobwebbing   Production of fine filaments instead of the normal atomized particles. A phenomenon observed during spray application characterized by the formation of web-like threads along with the usual droplets leaving the spray gun nozzle.

Coefficient of Thermal Expansion   A material characteristic used to predict the change in material dimensions as a function of temperature changes.

Cohesion   The degree of internal bonding of one substance to itself (see ADHESION).

Cold-Applied   Capable of being applied without heating. Cold-applied products are furnished in a liquid state, whereas hot-applied products are furnished as solids that must be heated to liquefy them.

Colloidal Dispersion   A mixture wherein a finely divided material is uniformly distributed within a liquid. LATEX (see also) emulsion is a colloidal dispersion of resin in water.

Color Stability   The ability to retain the original color without significant change over time.

Combustible   Capable of burning.

Comparator   An eye piece with magnification ranging from 4 to 12 power, with a scale used for measuring thickness.

Compatible Materials   Two or more substances that can be mixed, blended, or attached without separating, reacting, or affecting the materials adversely.

Compressive Strength   The stress at yield point or at 10 percent deformation (the latter is used in practice). Compressive strength is measured in kPa or psi.

Condensation   The action of vapor (or gas) converting into liquid.

Conditioning   The exposure to environmental conditions for a specified period of time or until a stipulated relation between material and the environment is reached.

Conductor Head (Lead)   A transition component between a through-wall scupper and downspout to collect and direct run-off water. (Also called the COLLECTOR BOX.)

Control Joint   See: AREA DIVIDER, EXPANSION JOINT.

Coping   The covering at the top of a wall or parapet designed to shed water.

Copolymer   A polymer consisting of molecules containing large numbers of units two or more chemically different types in irregular sequence.

Counterflashing   Formed metal or elastomeric sheeting secured on or into a wall, curb, pipe, roof-top unit, or other surface to cover and protect the upper edge of a base flashing and its associated fasteners.

Coverage   The unit quantity of material necessary to apply to achieve a desired thickness. Usually expressed in square meter (feet) per gallon or gallons per roofing square (hundred square feet).

Crazing, Craze Cracks   Fine, random cracks on the surface of a coating, often caused by shrinkage.

Cream Time   Time, measured in seconds at a given temperature, when the polyurethane components will begin to expand after being mixed.

Creep   (1) The permanent deformation of a material caused by a prolonged mechanical or thermal or loads, also (2) lateral movement of foam during application (lateral expansion).

Cricket   A relatively small, elevated area designed to facilitate the flow of water around an obstruction on a roof such as a chimney or a skylight.

Cure   The chemical and physical changes in a substance, which result in achieving the long-time characteristics of the substance.

Curtain Wall   A lightweight exterior wall system supporting no more than its own weight, the roof and floors being carried by an independent structural framework.

Deck   The structural surface to which the roofing or water proofing system (including insulation) is applied.

Deflection   The deviation of a structural element from its original shape or plan due to physical loading, temperature gradients, or movement of its support.

Degradation   The deterioration of a substance caused by contact with its environment (weathering).

Delamination   Separation or lack of adhesion between layers of polyurethane foam and/or coatings and/or substrate. May result in formation of a BLISTER.

Dew Point   The temperature at which a vapor begins to condense.

DFT   Dry film thickness.

Diffusion   The movement of water vapor from regions of high concentration (high water vapor pressure) toward regions of lower concentration (due to random thermal molecular motion).

Diisocyanate   An organic chemical compound having two reactive isocyanate (—N═C═O) groups; used in the production of polyurethane foams and coatings. (See MID also.)

Dimensional Stability   The ability of a material to retain its original size and shape (measured as a percent of original dimension). For polyurethane foam, dimensional stability is determined as a comparative characteristic under specified conditions of temperature and humidity.

Discoloration   Any change from the initial color. (See also: COLOR STABILITY.)

Drip Edge   A projecting element, shaped to throw off water and prevent its running down the face of the wall surface.

Dry   (verb) To change the physical state of a material by the loss of components through evaporation, absorption, oxidation, or a combination of these effects; (2) the absence of moisture.

Dry Bulb Temperature   The temperature of air as measured by an ordinary thermometer. (See also: AMBIENT TEMPERATURE.)

Dry Film Thickness   The thickness, expressed in mils, of an applied and cured coating or mastic.

Drying Time   For a coating, it is the time required for the loss of volatile components, so that the material will no longer be adversely affected by weather phenomena such as dew, rain, or freezing (cf. CURE TIME).

Durability   The ability to withstand physical, chemical, or environmental abuse over prolonged period.

Elastomer   A material which at room temperature is capable of being stretched repeatedly at least twice its original length (100 percent ELONGATION) and, upon release of stress, will return to its original dimensions.

Elongated Cells   Excessively large cells in foam or coating generally caused by off-ratio materials, moisture contamination, or excessive heat.

Elongation   The increase in length of a specimen at the instant that rupture occurs (expressed as a percent of the original length).

Emulsion   A COLLOIDAL DISPERSION of one liquid in another. (See: LATEX)

Epoxy   A class of synthetic, thermosetting resins which produce tough, hard, chemical-resistant coatings and adhesives.

Epoxy Coating   A coating based on cross-linking resins having the oxirane structure, noted for high mechanical strength, good adhesion, and resistance to solvents, acids, alkalines, and corrosion.

Exotherm   Heat generated by a chemical reaction.

Expansion Joint A joint designed to accommodate movement in the structure caused by variation of environmental and mechanical loads.

Fast Set A coating system with a very fast initial cure time, usually five seconds to one hour.

Feathered Edge The thin tapered outside edge of a polyurethane foam pass.

Felt Paper A building paper saturated with hot bitumen and rolled smooth, used in under roofing and siding materials as protection against moisture and air infiltration.

Filler An inert ingredient added to coating or polyurethane foam formulation to modify physical characteristics.

Film Thickness The thickness of a membrane or coating. Wet film thickness is the thickness of a coating as applied; dry film thickness is the thickness after curing. Film thickness is usually expressed in mm or mils (thousandths of an inch).

Fire Resistance The ability of a building material or component to resist the exposure to fire.

Fisheye Coating defect that manifests itself by the cracking of wet coating into a recognized pattern resembling small "dimples" or "fish eyes."

Flame Retardant A substance which is added to a coating or polyurethane foam formulation to reduce or retard its tendency to burn.

Flame Spread A relative measure of fire propagation rate; the flame spread of a tested material is rated relative to asbestos cement board (flame spread = 0) and red oak flooring (frame spread = 100).

Flammability Relative ability of a material to support combustion as expressed by its flash point.

Flash Point The lowest temperature of a material at which it gives off vapors sufficient to form an ignitable mixture with air near its surface.

Flashing The portion of a system used for preventing water ingress at terminations or vertical surfaces.

Fluorocarbons Components used as blowing agents.

Flutes The shaped lower section of metal decking that gives it added strength.

Flow Movement of an adhesive, coating, or sealant during the application process before set has occurred.

FM Factory Mutual. An independent testing agency, specialized in fire testing; see UL, and ULC.

Foam Stop   The roof edge treatment upon which polyurethane foam is terminated.

Frame Wall   Typically a platform frame in which vertical metal or wood wall structures (studs) extend between bottom plate rests on the subfloor of the storey and top plate supports floor joists of the next storey.

Friability   The tendency of a material or product to crumble or break into small pieces easily.

Granule   Size No. 11 ceramic aggregate embedded into wet coating over polyurethane foam for aesthetics, traction, and mechanical resistance.

Grit Blasting   Abrasive cleaning of surface by blasting with angular iron grit, aluminum oxides, or any crushed or irregular abrasive. The grit is projected onto the surface either mechanically or by means of compressed air.

Hardness   Ability of a coating film, as distinct from its substrate, to resist indentation or penetration by a hard object.

Heat Aging   Controlled exposure of materials to elevated temperatures for a period of time.

Heat Sink   A cold substrate which absorbs the SPC exothermic heat, slowing down the reaction and rise of the foam.

Hiding Power   The ability of a coating to hide or obscure a surface to which it has been uniformly applied.

Hydrophilic   Having an affinity or attraction for water.

Hydrophobic   Having no affinity for water; not compatible with water.

Hygroscopic   Attracting and absorbing atmospheric moisture.

IBC   International Building Code.

ICAA   International Contractors Association of America.

ICBO   International Conference of Building Officials, a model building code in the U.S.

Ignition Temperature   The minimum temperature to which a solid, liquid, or gas must be heated to initiate or cause self-sustained combustion independent of the heating element.

Impact Resistance   Ability to withstand mechanical or physical blows without the loss of protective properties. The impact resistance of the assembly is a function of all its components, not just the membrane itself.

Interlaminar Adhesion Adhesion between polyurethane foam passes or coating passes.

ISO Short form of ISOCYANATE.

Isocyanate A highly reactive organic chemical containing one or more isocyanate ($-N\!=\!C\!=\!O$) groups. A basic component in polyurethane foam chemical systems and some polyurethane coating systems.

***k*-Factor** Thermal conductivity of the material; expressed in $[W/(m\ K)$ (Btu in./(hr ft^2 °F)].

Knit, Pass or Lift Line Terms describing the adhesion place where one pass is sprayed over another.

Krebs Units (K.U.) A measurement of viscosity for materials that have the property of changing resistance to flow when under shear. Such materials are called thixotropic. Measuring is done with a Krebs/Stormer viscometer.

Latex A COLLOIDAL DISPERSION (see also) of a polymer or elastomer in water which coalesces into a film upon evaporation of the water.

Lift The sprayed polyurethane foam that results from a pass.

Mastic A coating material of relatively thick consistency.

Material Safety Data Sheet (MSDS) A standard formatted information sheet prepared by a material manufacturer which describes the potential hazards, physical properties, and procedures for safe use of a material.

Mechanical Damage Breaks or punctures to insulation and coating systems as a result of impact or abrasion.

Membrane Reinforcement Fabrics or fibers embedded in mastic or coating to provide strength and impact resistance.

Methylene Diphenyl Diisocyanate (MDI) Component A in polyurethane foam. An organic chemical compound having two reactive isocyanate ($-N\!=\!C\!=\!O$) groups. It is mixed with the B component to form polyurethane.

Mil One-thousandth of an inch; 0.001 in (0.025 mm). A unit used to measure coating thickness.

Mist Coat Very thin sprayed coat.

Mold Fungal growths often resulting in deterioration of organic materials, especially under damp conditions.

Monolithic Formed from or composed of a single material; seamless.

MSDS (See: MATERIAL SAFETY DATA SHEET)

Mud Cracking   The defect in an applied coating or mastic when it cracks into large segments or shrinks (also called ALLIGATORING). When the action is fine and incomplete, it is usually referred to as CHECKING.

Multiple Coat   Two or more layers of coating applied to a substrate.

NBC   National Building Code. The model building code in Canada.

Neoprene Rubber   A synthetic rubber having physical properties closely resembling those of natural rubber; made by polymerization of chloroprenes.

Night Seal   A material and/or method used to temporarily seal a membrane edge during construction to protect the roofing assembly in place from water penetration. May be removed when roofing application is resumed.

Non-Breathing Membrane   A membrane material which has a significantly greater resistance to the diffusion of water vapor than the other materials with which it is used.

Non-Flammable   Liquid having no measurable FLASH POINT.

NRCA   National Roofing Contractors Association.

Off-Ratio Foam   Polyurethane foam which has a lack of isocyanate or resin. Off-ratio foam will not exhibit the physical properties of normal foam.

Orange Peel Surface Texture   The surface texture of polyurethane foam resembling that of an orange peel.

Organic   Compounds containing carbon.

Overspray   (1) Airborne loss of SPF or coatings; (2) undesirable depositions of airborne spray loss.

Overspray Surface Texture   The surface shows a linear coarse textured pattern and/or a pebbled surface. This surface is generally downwind of the sprayed polyurethane path and is unacceptable for proper coating coverage and protection, if severe.

Oxidation   A combination of ultraviolet light and exposure to oxygen that causes substances to degrade.

Parapet   A wall or top portion of a wall extending above an attached horizontal surface such as a roof, terrace, or deck; often used to provide a safety barrier at roof edge.

Pass   The amount of coating or polyurethane foam applied by moving the gun from side to side and moving away from fresh material. A pass is delineated by its width, length and thickness.

Pass Lines   The overlapping of the polyurethane foam or coating as the newly applied material ties into the previous pass.

Peel Strength   The average force (or force per unit width) required to peel a membrane or other material from the substrate to which it has been bonded.

Peeling   Top coat inadequately bonded with undercoat resulting in partial delamination or detachment of final coat.

Perm   A unit of water vapor transmission defined as 1 grain of water vapor per square foot per hour per inch of mercury pressure difference.

Permeability   The rate at which water vapor will diffuse through a unit thickness and area of material, induced by a unit differential in water vapor pressure; units are usually $ng/(m^2 \cdot s \cdot Pa)$ or grains in./$(ft^2$ hr in) or Perm inch.

Permeance   The rate at which water vapor will diffuse through a unit area of material induced by a unit differential in water vapor pressure. Permeance values are reported for specific thickness (usually recommended application thickness). Units are usually $ng/(m^2$ s Pa) or Perm.

pH   A measure of acidity/alkalinity of aqueous mixtures. A measure of pH 7 is neutral, lower is more acidic, higher is more alkaline.

Pigment   Fine solid particles, dispersed in a coating to impart color.

Pinhole   A small hole in a mastic, coating, or polyurethane foam.

Plasticizer   A substance added to a plastic or coating to increase its flexibility or elongation.

Polymer   A substance consisting of chemical compounds with high molecular weight characterized by chains of repeating simpler units.

Polyol   Polyol is the main ingredient of the resin component which reacts with the isocyanate to form polyurethane.

Popcorn Surface Texture   A polyurethane foam surface texture where valleys form sharp angles. This surface is unacceptable for uniform coating application.

Positive Drainage   The drainage condition in which there is no consequential standing water on the roof 48 hours after a rainfall, during ambient drying conditions.

Pot Life   The period of time during which a multi-component or catalyzed material remains suitable for application after being mixed.

Primer   The first layer of coating applied to a surface to improve the adhesion of subsequently applied materials or to inhibit corrosion.

Proportioner   The basic pumping unit for spraying polyurethane foam or two component coatings systems. Consists of two positive displacement pumps designed to dispense two components at a precisely controlled ratio.

Psychrometer   A device for measuring ambient humidity by employing a dry bulb thermometer and a wet bulb thermometer.

Psychrometric Chart   A diagram relating the properties of humid air with temperature.

Purge   To cleanse or remove liquid materials from equipment or shoes.

R-Value (Thermal Resistance)   The resistance of the heat transfer through a material (equal to the thickness of the material divided by the k factor). Units are $(m^2 K)/W$ or $(hr\ ft^2\ °F/Btu)$.

RCI   Roof Consultants Institute.

Recovering   The process of installing a new roofing system over an existing roofing system.

Relative Humidity   The ratio of absolute humidity to saturation humidity, expressed as a percent.

Re-Roofing   Either RE-COVERING or REPLACEMENT of a roofing system.

Resin   (1) General term applied to a wide variety of more or less transparent and fusible products, which may be natural or synthetic. Higher molecular weight synthetic resins are referred to as polymers. (2) Any polymer that is a basic material for coatings and plastics. (3) Component B in SPF. This component contains a catalyst, blowing agent, fire retardants, surfactants and polyol; it is mixed with the A component to form polyurethane.

Retrofit   The modification of an existing building or facility to include new systems or components.

Roof Slope   The roof angle measured in the number of inches of vertical rise in a horizontal length of 12″.

Rust Blush   The earliest stage of rusting characterized by an orange or red color. Occurs frequently on freshly sandblasted steel if allowed to stand too long before coating.

Sag   Undesirable excessive flow in material after application to a surface.

Saturation Humidity   The maximum concentration of a given temperature before vapor condensation occurs.

SBCCI   Southern Building Code Congress International. One of the model building codes in the U.S.

Scarf To remove the surface or coating from polyurethane foam by cutting, grinding, or other mechanical means. Synonymous with scarify.

Scrim A woven, non-woven, or knitted fabric, composed of continuous strands of material used for reinforcing or strengthening membranes. Scrim may be incorporated into a membrane by the laminating or coating process.

Scupper An opening in a parapet wall allowing runoff water to exit a roof.

SCV Solid content by volume.

Sealant Any material used to close up cracks or joints to protect against rain water leaks.

Service Temperature Limits The maximum temperature at which a coating, polyurethane foam, or other material will perform satisfactorily.

Set To convert into a fixed or hardened state by chemical or physical action.

Shelf (Storage) Life The period of time within which components of a product (SPF of coating) remain suitable for use.

Skinning The formation of a dense film on the surface of a liquid coating or mastic.

Slit Samples Small cut samples [e.g., 50 mm (2″) long, 12 mm (1/2″) deep] used for evaluation.

Smooth Surface Texture The surface shows spray undulation and is ideal for receiving a protective coating.

Solids Content The percentage of non-volatile matter in a coating or mastic formulation; may be expressed as a volume or weight percent.

Solvent A liquid that dissolves other substances. (See: THINNER.)

SPI/SPFD The Society of the Plastics Industry/Sprayed Polyurethane Foam Division.

Sprayed Polyurethane Foam (SPF) (1) A foamed plastic material, formed by spraying two components, isocyanate ("A" component) and a resin ("B" component); (2) a generic term representing either sprayed, froth, poured, two-component (catalytic) or one component (moisture) cured, field fabricated polyurethane foam.

Square (Roofing Square) The roofing area equal to 100 square feet.

Stress An applied force, which tends to deform a body. May either be a tensile stress (pulling or stretching force), compressive stress (pushing or compacting force) or shear stress (opposite but offset parallel forces tending to produce a sliding motion).

Stress Cracking   Cracking caused by presence of long-term stresses.

Substrate   The surface to which polyurethane foam or coating is applied.

Surface Erosion   The wearing away of a surface due to abrasion, dissolution, or weathering.

Surface Texture   The resulting surface from the final pass of SPF. The following terms are used to describe the types of SPF surfaces: smooth, orange peel, coarse orange peel, verge of popcorn, popcorn, treebark, and oversprayed.

Surfacing   The top layers of a roof covering, specified or designed to protect the underlying roofing from direct exposure to the weather.

Surfactant   Short for "surface active agent" that is used to alter the surface tension of liquids. An ingredient in polyurethane foam formulations to aid in mixing and controlling cell size.

Tack-Free   A curing phase of polyurethane foam or coating wherein the material is no longer sticky.

Tensile Strength   The tensile (pulling or stretching) force necessary to rupture a material sample divided by the sample's original cross-sectional area. Units are usually kPa or psi.

Termination   The treatment or method of anchoring and/or sealing the free edges of the membrane in roofing or waterproofing systems.

Thermal Barrier   A material applied over polyurethane foam designed to slow the temperature rise of the foam during a fire situation and delay its involvement in the fire. Thermal barriers for use with polyurethane foam must have a time rating not less than 15 minutes.

Thermal Bridge   A thermally conductive material which penetrates or bypasses an insulation system, such as a metal fastener or stud.

Thermal Conductivity Coefficient (*k*-factor)   The quantity of heat conducted under steady state conditions, in one second (hour), through 1 square meter (foot), under a difference of 1 degree Celsius (Fahrenheit) across the two parallel surfaces of a homogeneous material with 1 meter (inch) thickness.

Thermal Resistance (*R*-value)   The resistance to heat transfer through a material (equal to the thickness of the material divided by the *k*-factor). Units are $(m^2\,K)/W$ or $(hr\,ft^2\,°F/Btu)$.

Thermal Shock   The stress producing phenomenon resulting from sudden temperature drops in a roof membrane, for example, a rain shower following brilliant sunshine.

Thermoplastic   A polymeric material that repeatedly softens upon heating and hardens upon cooling.

Thermoset   A polymeric material whose physical properties are relatively unaffected by modest changes in temperature; when heated thermoset material will degrade rather than melt.

Thinner   A liquid used to reduce the viscosity of coatings or mastics. Thinners evaporate during the curing process. Thinners may be used as solvents for clean-up of equipment.

Thixotropic   Having the property of decreasing viscosity with increasing shear stress. A coating is thixotropic if it thins with stirring or pumping but thickens back up when movement ceases.

Tie in Lines   The starting or stopping point at which new foam is applied to foam which had been sprayed earlier.

Tint   A color produced by the introduction of small amounts of a colored pigment.

Toxicity   The quality, property, or degree of being poisonous or toxic.

Treebark Surface Texture   A rolling foam surface texture where the values form sharp angles generally caused by spraying foam at an angle. This surface is unacceptable for coating or application.

Two-Part System   A coating or polyurethane foam formed by the mixing and the reaction of two different materials.

UBC   Uniform Building Code. Model building code generated by ICBO.

UL   Underwriter's Laboratory. An independent testing agency, specialized in fire testing.

ULC   Underwriter's Laboratory of Canada. An independent testing agency, specialized in fire testing.

Ultra-Violet Radiation (UV)   Electromagnetic radiation beyond the visible spectrum at its violet end. Invisible, high-energy sunlight, which degrades many organic materials.

Underlayment   A material that is laid down as a substrate for the sprayed polyurethane foam (1) to make the surface smooth or to give a specific rating for interior fire exposures, or (2) to bridge the differential movements of the deck.

Urethane Coatings   A one or two part coating that contains polyisocy-anate monomer and a hydroxyl containing material, which react during cure to form polyurethane.

UV   Abbreviation for "ultra-violet" (see also: ULTRAVIOLET RADIATION).

U-**Value**   Overall thermal conductance. *U*-value is equal to the inverse of the sum of the *R*-values in a system.

Vapor Retarder (Barrier)   A film, coating, sheet, or other building com-ponent which reduces the migration of water vapor to the building compo-nents with which it is used.

Verge of Popcorn Surface Texture   The verge of popcorn surface tex-ture is the roughest texture suitable for receiving the protective coating on a sprayed polyurethane foam roof. This surface is acceptable for receiving a protective coating only because of the relatively curved valleys. However, the surface is considered undesirable because of the additional amount of coating material required to protect the surface properly.

Viscosity   The thickness or resistance to flow of a liquid. Viscosity gener-ally decreases as temperature increases.

Volatile Organic Compounds (VOC)   Any compound containing car-bon and hydrogen or containing carbon and hydrogen in combination with other elements that has a vapor pressure of 10 kPa (1.5 pounds per square inch) absolute or greater under actual storage conditions.

Walkways   High traffic and high service areas on a rooftop, particularly those with vents, hatches, and heavy duty air conditioning units, that have been reinforced with extra coating and granules to prevent damage to the polyurethane foam system.

Wet Bulb Temperature The temperature of air as registered by a ther-mometer whose bulb is covered by a water wetted wick. Units are °C or °F.

Wet Film Thickness   The thickness, expressed in mm or MILS (see also), of a coating or mastic as applied but not cured. For comparison, see DRY FILM THICKNESS.

Windscreen   A device to minimize the effects of wind on coating or polyurethane foam application.

Wind Uplift   The force caused by the deflection of wind at roof edges, roof peaks, or obstructions, causing a drop in air pressure immediately above the roof surface.

Standards for Surface Texture
Used by the SPF Industry

Figures A1–A8 show standards of SPF surface texture as agreed by the industry. Acceptable surfaces range from smooth [Figure (A1)], orange peel [Figure (A2)] and coarse orange peel [Figure (A3)]. Verge of popcorn [Figure (A4)] is marginally acceptable. Popcorn [Figure (A5)], Treebark [Figure (A6)] are not acceptable for roofing applications and Pinholes [Figure (A7)] or Rippling [Figure (A8)] indicate problems which must be corrected.

FIGURE A1. Polyurethane foam texture, smooth.

FIGURE A2. Polyurethane foam texture, orange peel.

283

FIGURE A3. Polyurethane foam texture, coarse orange peel.

FIGURE A4. Polyurethane foam texture, rippling–verge of popcorn.

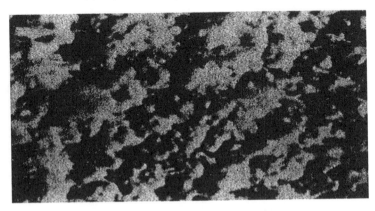

FIGURE A5. Polyurethane foam texture, popcorn.

FIGURE A6. Polyurethane foam texture, treebark.

FIGURE A7. Machine problems, which must be corrected (lack of component A—isocy-anate).

FIGURE A8. Machine problems, which must be corrected (lack of component B—polyol).

Material Specification—SPF for New and Remedial Roofs

1. MATERIAL SPECIFICATION, PART 1—GENERAL CONSIDERATIONS

1.1 Scope of Work
1.2 Related Work Specified Elsewhere
1.3 Quality Assurance
1.4 Submittals
1.5 Materials, Delivery, and Storage
1.6 Environmental Conditions
1.7 Sequencing and Scheduling
1.8 Warranty
1.9 Safety Requirements

2. MATERIAL SPECIFICATION PART 2—PRODUCTS

2.1 SPF
2.2 Protective Coating
2.3 Accessories and Miscellaneous Materials

3. MATERIAL SPECIFICATION PART 3—EXECUTION

3.1 General
3.2 Surface Preperation and Priming
3.3 SPF Application
3.4 Protective Coating Application
3.5 Granule Application
3.6 Walkways
3.7 Safety Requirements

1. MATERIAL SPECIFICATION, PART 1—GENERAL

This specification discusses the seamless SPF with a protective coating for use as an insulated roofing system either for new or retrofit roofing. The contractor, SPF system manufacturer, and code agency can assist the designer, as each project may need to be assessed individually.

1.1 Scope of Work

Furnish all labor, materials, tools, and equipment necessary for the application of a SPF roofing system, including accessory items, subject to the general provisions of the contract.

1.2 Related Work Specified Elsewhere

A. Cast-In-Place Concrete   Section 03300
B. Metal Decking   Section 05300
C. Rough Carpentry   Section 06100
D. Insulation   Section 07200
E. Membrane Roofing   Section 07500
F. Flashing and Sheet Metal   Section 07600
G. Roof Specialties and Accessories Section 07700
H. Skylights   Section 07800
 I. Mechanical   Division 15
 J. Electrical   Division 16

1.3 Quality Assurance

A. *Contractor Qualifications*[72]: The proposed contractor should provide information concerning projects similar to the one proposed, including location and person to be contacted. Some manufacturers of SPF systems and/or protective coatings have specific approval/training and/or licensing programs for SPF contractors.
B. *Manufacturer Qualifications:* SPF and protective coating manufacturers shall show evidence of sufficient financial resources and manufacturing facilities to furnish materials on this project. References shall be required,

---

[72] Accreditation Programs developed for the purpose of improvement of quality assurance in the application of SPF and coatings through education and good business practices are a service to the owner and/or specifier as well as to those involved at all levels of the industry. Individual accreditation attests that required courses and field inspection requirements have been successfully completed. Some building codes specifically require that an SPF contractor be licensed by the accredited training program. A list of accredited individuals and firms is available from SPFD/SPI and CUFCA.

sufficient project lists, warranties and code approvals shall be submitted for verification.

C. *Inspections:* The SPF and protective coating manufacturers may provide qualified representatives to monitor and inspect the installation of their products. Third party inspection of the installation is recommended. A list of SPFD/SPI or CUFCA inspectors is available.

1.4 Submittals

A. Manufacturers are to provide published data sheets or letter of certification that their products comply with the materials specified. This is to include primers (if required), SPF, and protective coatings.

B. Shop drawings on sheet metal, accessories, or other fabricated items.

C. Instructions for SPF installation, if different from SPFD or CUFCA manuals.

D. Contractor/applicator certification from SPF supplier and/or protective coatings manufacturers and evidence of contractor/applicator qualification and experience (see Section 1.3A).

E. A specimen copy of the applicable warranty for the project (see Section 1.3B).

F. Approval and information guides for either applicable local or national codes and/or insurance acceptability, if required.

G. Safety and handling instructions for storage, handling, and use of the materials to include appropriate Materials Safety Data Sheets (MSDS).

H. Field quality assurance procedures to be utilized by the contractor/applicator to insure proper preparation and installation of SPF (see Chapter 6 of the previous issue of the journal) and protective coatings, detail work, and follow-up inspection.

I. Spray Polyurethane Foam Division of the SPI Inc., the SPF surface visual guide (photos as shown in Appendix A3).

1.5 Materials, Delivery, and Storage

A. Materials shall be delivered in the manufacturer's original, tightly sealed containers or unopened packages, all clearly labeled with the manufacturer's name, product identification, safety information, and batch or lot numbers where appropriate. Where materials are covered by a referenced specification, the labels shall bear the specification number, type, and class, as applicable.

B. Containers shall be stored out of the weather and direct sunshine where the temperatures are within the limits specified by the manufacturer.

C. All materials shall be stored in compliance with local fire and safety requirements.

1.6 Environmental Conditions

A. The SPF application shall not proceed during periods of inclement weather. Do not apply the SPF below the temperature and/or above humidity specified by the manufacturer for ambient air and substrate.

B. Do not apply protective coatings when there is ice, frost, surface moisture, or visible dampness present on the surface to be coated. Prior to applying the coatings, check the SPF to insure that the surface is dry. Apply protective coatings in accordance with the coatings' manufacturer's application instructions.

C. Wind barriers may be used if wind conditions could affect the quality of the SPF or protective coating installation.

1.7 Sequencing and Scheduling

In new construction, the SPF is installed when the deck, parapet walls, rough openings, and curbs are completed. The type of skylight used will determine when skylights should be installed. Plumbing vents, drains, and electrical penetrations should all be in place. There should not be any other trades-people working on the roof when the SPF and coating are being installed.

1.8 Warranty

Warranty agreements vary in duration and content. If a warranty is desired, it is suggested that parameters be established as a prerequisite to the contract.

1.9 Safety Requirements

A. See SPI Bulletin AX-119, "MDI-Based Polyurethane Foam Systems: Guidelines for Safe Handling and Disposal."

B. Refer to appropriate Materials Safety Data Sheets (MSDS) for additional safety information.

2. MATERIAL SPECIFICATION, PART 2—PRODUCTS

2.1 SPF

A. The SPF to be applied shall be a HD-type, two-component SPF system made by combining an isocyanate (A) component with a polyol (B) component and shall possess the following physical characteristics:

Properties	ASTM Test	Value/Units
Minimum overall density	D 1622 or buoyancy[73]	45 kg/m³ or 2.8 lbs/ft³
Minimum compressive resistance[74]	C 165 to D 1621	280 kPa or 40 psi
Thermal resistance per inch, aged SPF 1-1/2 inch thick layer or more	C 518 and LTTR determined with C 1303	1.05 m K/W or 6.0 (ft² hr °F)/BTU in
Flame spread index (FSI)[75]	E-84	< 75
Smoke development index[75]	E-84	< 450

B. SPF Primers shall be used as recommended by the manufacturer of the SPF.

C. Fire Safety Requirements—See SPI Bulletin AX-119, "MDI-Based Polyurethane Foam Systems: Guidelines for Safe Handling and Disposal."

2.2 Protective Coatings

A prerequisite in SPF roofing systems is the protective coating. Normally, the foam is protected from ultraviolet (UV) radiation and excessive moisture by one of the following protective layers:

1. Acrylics
2. Butyls
3. Chlorinated synthetic rubbers[76]
4. Silicones
5. Elastomeric urethanes
6. Modified asphalts
7. Gravel layer[77]

Performance objectives for elastomeric coatings were discussed in Chapter 4 of this monograph. For further information concerning protective coatings

---

[73]A field density test specific in Chapter 6 of this monograph is equivalent to the ASTM test in question.

[74]Compressive resistance is a compressive strength at yield point or at 10% deformation, whichever comes first.

[75]This standard is used to measure the material response to flame under controlled laboratory conditions. The FSI is not intended to describe hazards presented by the material under actual fire conditions.

[76]Often known under the registred trademarks of E. I. DuPont de Nemours Co., "Hypalon" or "Neoprene" coatings.

[77]As this protective layer does not address the moisture control aspects of system performance, its use may be restricted to specific climatic and drainage conditions.

refer to "A Guide for Selection of Protective Coatings over Sprayed Polyurethane Foam Roofing Systems" [SPI Stock Number AY 102].

2.3 Accessories and Miscellaneous Materials

A. Flashings and waterproof coverings for expansion joints shall be compatible with the specified SPF/coating and shall be as recommended by the
manufacturers of the systems used.
B. Miscellaneous materials such as adhesives, caulking compounds, metal,
vents, and drains shall be part of the roof system and shall be those recommended by the systems manufacturer.
C. Granules (optional), when used, shall be of those as recommended by the
coating manufacturer.
D. Underlay board, if required over metal decks, should be fastened to
achieve the required wind uplift resistance.

3. MATERIAL SPECIFICATION, PART 3—EXECUTION

3.1 General

Guidelines outlined by the manufacturer issuing the warranty shall determine the final specification.

3.2 Surface Preparation and Priming

A. Built-up Roof (Retrofit)
  1. All loose gravel, dust and residue shall be removed using power vac-
     uum equipment, power sweeper, air blowing, or other suitable means.
  2. The roof shall be thoroughly inspected or tested to determine if mois-
     ture is present within the roof assembly.
  3. The existing roof shall be thoroughly inspected for adhesion between
     felts, insulation, and deck. Areas of poor adhesion should be fastened.
     Blisters, buckles, wrinkles, and fishmouths shall be cut out and/or fas-
     tened.
  4. All soft mastic or other materials that impede polyurethane adhesion
     shall be removed.
  5. Remove or refasten all loose base flashings, counterflashings, and
     gravel stops as required.
  6. The need may exist for structural design analysis to determine expan-
     sion joint requirements. Existing expansion joints should be inspected
     and repaired if necessary.

7. Lightning rods shall be masked prior to foaming. Lightning rod cables shall not be embedded in the SPF and should be removed prior to foaming. Electrical and mechanical conduits should be relocated or raised above the finished roof surface.

B. Metal Deck

1. The metal roof deck shall be constructed of a minimum 22 gauge steel. Construction shall conform to local building codes.

2. Ferrous Metal—sandblast iron and steel surfaces which are not primed, shop painted, or otherwise protected in accordance with SSPC SP-6, Commercial Blast Cleaning. Remove loose rust and unsound primer from shop primed iron and steel surfaces by scraping or wire brushing.

3. Non-Ferrous Metal—clean galvanized metal, aluminum, and stainless steel surfaces as recommended by the manufacturer issuing the warranty.

4. If the metal surface is free of rust or chalking paint, it can be cleaned with compressed air jet, vacuum equipment, and hand or a power broom. Grease, oil, or other contaminants should be removed with proper cleaning solutions.

5. Fluted metal decks require a suitable method of covering or filling the flutes prior to SPF application. Flutes may be covered with mechanically fastened board stock, open-weave mesh fabric, or filled with pre-cut foam board or SPF.

C. Concrete

1. Remove loose dirt, dust and debris by using compressed air, vacuum equipment, or a broom. Oil, grease, form release agents, or other contaminants shall be removed with proper cleaning solutions.

2. All joint openings in concrete decks that exceed 1/4 inch shall be grouted or caulked prior to application of SPF.

3. Priming is required on concrete surfaces, and it is recommended that poured concrete decks be permitted to cure for twenty-eight (28) days prior to the application of primer/SPF.

4. SPF is not recommended for lightweight or insulating concrete unless tests have been made to determine that the adhesion is adequate. Otherwise, a mechanically fastened, underlay board is used.

D. Wood

1. Plywood shall be exterior grade not less than 1/2 inch thick, nailed firmly in place. Attachment must meet building code requirements for resistance to wind uplift.

2. Plywood shall contain no more than 18 percent water.

3. All untreated and unpainted surfaces shall be primed with an exterior grade primer.

4. Plywood joints wider than 1/4 inch shall be taped or filled with a suitable sealant.
5. Deck shall be free of loose dirt, grease, oil or other contaminants prior to priming or foam application. Remove loose dirt or debris by use of compressed air, vacuum, or broom. No washing shall be permitted.
6. Tongue and Groove, Sheathing, Planking: Large frequency of joints, possibility of variable openings and effects of shrinking, necessitate that a minimum 1/4 inch thick exterior grade plywood or another suitable covering be used as an overlay board.

E. Other surfaces (i.e., gypsum board, polyisocyanurate board)
   1. These materials are generally used over fluted metal decks and must be fastened to yield the necessary strength for the wind uplift.
   2. Boards shall be firmly butted together along all edges without gaps or openings. Joints exceeding 1/4 inch shall be caulked with a suitable sealant material.
   3. Moisture exposure will damage these materials and may be cause for replacement. Therefore, a special care must be taken to prevent them from getting wet: in the storage, on the job site, and after installation, but prior to being protected by the SPF.
   4. Remove loose dirt and debris by using compressed air, vacuum, or a light broom. No power broom is permitted due to possibility of damage.
   5. The installed materials shall be protected from spills of contaminants such as oil, grease, solvents, etc., as these materials cause soiling that cannot be readily removed from the board surfaces.

3.3 SPF Application

A. Inspection
   1. Prior to application of the foam, the surface shall be inspected to insure that conditions required by section 3.2 have been met.
   2. Substrate shall have sufficient slope to eliminate ponding of water. If the substrate does not have sufficient slope, then, ponding water must be eliminated by building in slope by the SPF application and proper placement of drains.
   3. The SPF application shall not proceed during periods of inclement weather. The applicator shall not apply the SPF below the temperature and/or humidity specified by the manufacturer for ambient air and substrate. Wind barriers may be used if wind conditions could affect the quality of installation.

B. Application
   1. The SPF shall be applied in accordance with the manufacturer's specification and instructions.

2. Areas to be built up to remove ponding water are to be filled in with SPF before the specified thickness of SPF is applied to the entire roof surface (see Section 3.2A).
3. The SPF must be applied in a minimal pass thickness of 1/2 inch.
4. SPF thickness shall be a minimum of one inch (recommended not less than 1½ inch). The SPF shall be applied uniformly over the entire surface with a tolerance of plus 1/4″ per inch of thickness minus 0″, except where variations are required to insure proper drainage or to complete a feathered edge.
5. The SPF shall be uniformly terminated a minimum of four (4) inches above the roof line at all penetrations (except drains, parapet walls, or building junctions). Foamed in place cants shall be smooth and uniform to allow positive drainage.
6. Detailing skylights is particularly important in that the SPF must be terminated below existing weep holes.
7. The SPF surface shall be allowed to cure sufficiently. The full thickness of SPF in any area shall be completed prior to the end of each day. If due to weather conditions more than 24 hours elapse between SPF and coating application, the SPF shall be inspected for UV degradation, oxidation or contamination. If any of the above conditions exist, the surface shall be prepared in conformity with the recommendations of the manufacturer issuing the warranty.

C. Surface Finish
1. The final SPF surface shall be "smooth, orange peel, coarse orange peel, or verge of popcorn." SPF surfaces termed "popcorn" or "treebark" are not acceptable. SPF in these areas shall be removed and re-foamed.
2. Any damage or defects to the SPF surface shall be repaired prior to the protective coating application.
3. The SPF surface shall be free of moisture, frost, dust, debris, oils, tars, grease, or other materials that will impair adhesion of the protective coating.

3.4 Protective Coating Application

A. Inspection

Prior to the application of the protective coating the SPF shall be inspected for suitability of base coat application as per Section 3.3. The SPF shall be clean, dry, and sound.

B. Application of the base coat
a. The base coat shall be applied the same day as the SPF application when possible. In no case shall less than two hours elapse between application of the SPF and application of the base coat. If more than 24

hours elapse prior to the application of base coat, the SPF shall be inspected for UV degradation.

    b. The SPF shall be free of dust, dirt, contaminants, and moisture before application of the base coat.

    c. The base coat shall be applied at a uniform thickness with the rate of application being governed by the SPF surface texture. Coatings shall be applied at such a rate as to give the minimum DFT specified by the protective coating manufacturer.

    d. The cured base coat shall be inspected for pinholes, uncured areas, or other defects. Any defects should be repaired prior to subsequent applications. The base coat shall be free of dirt, dust, water, or other contaminants before application of the top coat.

    e. The coating application shall not proceed during periods of inclement weather. The applicator shall not apply the protective coating below the temperature and/or above the humidity specified by the manufacturer for ambient air and substrate. Wind barriers may be used if wind conditions could affect the quality of installation.

C. Application and inspection of the top coat and/or subsequent coating

    a. Subsequent coating should be applied in a timely manner to insure proper adhesion between coats. Surface texture of SPF will affect DFT; additional material may be required in areas of coarse surface texture.

    b. The cured DFT of the finished multiple coat application shall be checked by taking slit samples and examining under magnification. Areas that are found to have less than the specified thickness shall be provided with additional coating.

3.5 Granule Application (Optional)

When used, granules shall be of the size and type as recommended by the coating manufacturer.

3.6 Walkways

Walkways may be installed for heavy traffic areas and around frequently serviced roof top units. Breathable walk pads should be as recommended by the coating manufacturer.

3.7 Safety Requirements

A. See SPI Bulletin AX-119, "MDI-Based Polyurethane Foam Systems: Guidelines for Safe Handling and Disposal."

B. Before starting to apply SPF or coating, all HVAC equipment on the roof must be turned off. Any other potential sources of air entry into the building envelope must be sealed.

Closing Note: This guide is to assist the specifier in design of a long-lasting SPF and coating roofing system. Nevertheless, it is the responsibility of the specifier to ensure that the manufacturer's specific recommendations are considered.

Material Specification—Bead-Applied Sealing Foams (BSF)

1. GENERAL CONSIDERATIONS

1. The performance of a SPF can be affected by all the component parts of a wall structure, as well as the service and climatic conditions inside and outside the structure. Proper design and specification review, contractor and material selection, coupled with the compatibility of the various components of a wall structure are necessary.

2. Consult with the respective material suppliers and the successful contractor to receive written confirmation of their agreement to all facets of the air sealing foams, including, but not limited to, material selection, compatibility, flashing details, etc.

3. Under normal heating and cooling conditions, vapor retarder may be required on the warm side of LD and BSF types of SPF, even though this requirement does not apply to MD- or HD-types of SPF.

4. Conform to sections of Division 1[78] as applicable. Amend to suit specific project and identify all sections affecting this section.

1.1 Quality Assurance

A. *Contractor Qualifications*[79]*:* The proposed contractor should provide information concerning projects similar to the one proposed, including location and person to be contacted. Some manufacturers have specific approval/training and/or licensing programs for SPF contractors.

---

[78]Note for specifiers: Division 1 should cover reference to form of Contract, Supplementary General Conditions and the following: code, standards, coordination, material storage and handling, waste disposal, temporary cleaning, manufacturer's directions, submittals, quality control, etc.

[79]Accreditation Programs developed for the purpose of improvement of quality assurance in the application of SPF and coatings through education and good business practices are a service to the owner and/or specifier as well as to those involved at all levels of the industry. Individual accreditation attests that required courses and field inspection requirements have been successfully completed. Some building codes specifically require that an SPF contractor be licensed by the accredited training program. A list of accredited individuals and firms is available from SPFD/SPI and CUFCA.

B. *Manufacturer Qualifications:* The manufacturer shall show evidence of sufficient financial resources and manufacturing facilities to furnish materials on this project. References and project lists may be required, warranties and code approvals shall be submitted.

C. *Inspections:* The polyurethane foam and protective coating manufacturers may provide qualified representatives to monitor and inspect the installation of their products. Third party inspection of the installation is recommended. A list of SPFD/SPI or CUFCA inspectors is available.

1.2 Submittals

A. Manufacturers are to provide published data sheets or letter of certification that their products comply with the materials specified.

B. Shop drawings on sheet metal, accessories, or other fabricated items.

C. Instructions for the SPF installation, if different from SPFD or CUFCA manuals.

D. Contractor certification and evidence of contractor/applicator qualification and experience.

E. A specimen copy of the applicable warranty for the project.

F. Approval and information guides for applicable either local, or national codes, and/or insurance acceptability, if required.

G. Safety and handling instructions for storage, handling, and use of the materials to include appropriate Materials Safety Data Sheets (MSDS).

H. Field quality assurance procedures to be utilized by the contractor/applicator to insure proper preparation and installation of SPF (see Chapter 6).

1.3 Materials, Delivery and Storage

A. Materials shall be delivered in the manufacturer's original, tightly sealed containers or unopened packages, all clearly labeled with the manufacturer's name, product identification, safety information, and batch or lot numbers where appropriate. Where materials are covered by a referenced specification, the labels shall bear the specification number, type, and class, as applicable.

B. Containers shall be stored out of the weather and direct sunshine where the temperatures are within the limits specified by the manufacturer.

C. All materials shall be stored in compliance with local fire and safety requirements.

1.4 Environmental Conditions

A. The BSF–SPF applications shall not proceed during periods of inclement weather. Do not apply the polyurethane foam below the temperature and/or above humidity specified by the manufacturer for ambient air and substrate.

1.5 Related Work

1. Concrete: Section 03300, Cast-in-Place Concrete
2. Structural Steel and Steel Joists: Section 05120
3. Metal Decking: Section 05300, Steel Deck
4. Masonry: Section 04200
5. Firestopping: Section 07270
6. Insulation: Section 07210
7. Drywall: Section 09250, Gypsum Wallboard
8. Caulking: Section 07900, Sealants

1.6 System Description

Seal all joints where air leakage can occur except those joints specified to be sealed in other Section of Works. Included, but not limited thereto, are the following:

- interior joint at sill plate
- joints between door and window frames and rough openings
- around electrical outlets, piping, ducts and conduit which penetrate the vapor barrier
- joints shown on drawings and/or schedules to be sealed as part of the air retarder

Specifier Note: Amend to suit project. Joints greater than 100 mm should be sealed with WAM membrane under appropriate section (see Appendix 6).

1.7 Warranty

1. Warrant work of this section against defects or deficiencies for a period of two years. Work is certified as substantially performed in accordance with General Conditions of the Contract.
2. Promptly correct, at own expense, defects or deficiencies, which become apparent within warranty period.

2. MATERIALS

BSF is a predominantly closed cell, either a single component liquid system dispensed from aerosol cylinder, or a two-component, froth polyurethane foam, dispensed from pressurized cylinders. The latter has a minimum density of 27 kg/m³ (1.7 lb/ft³) and its compressive strength at 10% deflection is equal to 14 psi.

3. EXECUTION

3.1 Surface Preparation

1. Examine areas before applying the BSF-SPF to ensure configuration, surfaces, dimensions and environment are suitable for foam insulation and services; and that execution of installation and its performance will not be adversely affected.

Surfaces to receive the SPF shall be frost-free and not coated with releasing agents or other coating affecting bond. Report in writing defects in such work including adverse conditions. Do not work until such conditions are rectified. Application of work of this section shall be deemed acceptance of existing work and existing conditions.

2. Prepare surface by brushing, scrubbing, scraping, or grinding to remove loose mortar, dust, oil, grease, oxidation, mill scale and other materials which will affect adhesion and integrity of foam insulation. Wipe down metal surfaces to remove release agents or other non-compatible coatings, using clean cellulose sponges or rags soaked in solvent compatible with foam insulation. Ensure surfaces are dry before proceeding.

3.2 Application

1. Apply BSF in strict accordance with manufacturer's written instructions, specifications, or recommendations.
2. Fill joints with the foam sealant making allowance for post-expansion of foam. Use manufacturer recommended dispensing gun and follow sealant manufacturer's written instructions.

   Specifier Note: Show or specify locations of BSF.

3. Finish joints shall be free of air pockets and embedded foreign materials. Cut back excess BSF after cutting flush with surrounding surfaces unless otherwise directed and/or detailed.
4. Take care not to mark or cover surfaces beyond joints. Use masking tape if necessary.
5. Where BSF is to be faced by other caulking materials for finishing pur-

poses, ensure that the BSF is cut back or recessed to a sufficient depth for such caulking.

6. Remove BSF smears, droppings, etc., beyond joint, and masking tape, immediately after BSF has cured to hard surface film. Clean and make good surfaces soiled or damaged by work of this section. Consult with sections where work is soiled before cleaning to ensure methods used will not damage their work. Damage caused to work of other sections shall be made good by section whose work is damaged but at expense of section causing damage.

7. Finished SPF shall be free of voids and embedded foreign materials; and to uniform minimum thickness shown or specified.

8. Do not allow foam insulation to cover or spot surfaces beyond surfaces to be sprayed. Use masking materials if necessary.

9. Remove over-spray and masking materials immediately after foam has cured.

Material Specification—Water, Air, and Movement (WAM) Membranes

1. GENERAL CONSIDERATIONS

1. Normally, the mechanical fatigue of SPF is unknown, and underlay boards are used when a crack bridging performance is required. For instance, when mechanical or hygric-based movements are expected in a wood plank roof-deck, an overlay board is recommended (see Chapter 9). Similarly, a strip of WAM membrane is to be placed in all places, such as joints and junctions, where movements between component parts of a structure are expected.
2. Consideration must be given to the substrate movements, as they may be affected by the service and climatic conditions inside and outside the structure. When maintaining the continuity of the SPF layer is necessary, or existing cracks/openings are expected to move more than 3 percent of the SPF layer thickness, use of WAM membrane is also recommended.
3. The WAM membranes may also be used for flashing details, details related to water drainage or air sealing. Consult with the successful contractor, or the respective material supplier, to ensure the material compatibility.
4. If the SPF layer is applied over a 100 mm-wide joint (or wider than 100 m), this joint should be sealed with the WAM membrane, independently of its predicted movement.

1.1 Quality Assurance

A. *Contractor Qualifications*[80]: The proposed contractor should provide in-

---

[80]Accreditation Programs developed for the purpose of improvement of quality assurance in the application of SPF and coatings through education and good business practices are a service to the owner and/or specifier as well as to those involved at all levels of the industry. Individual accreditation attests that required courses and field inspection requirements have been successfully completed. Some building codes specifically require that an SPF contractor be licensed by the accredited training program. A list of accredited individuals and firms is available from SPFD/SPI and CUFCA.

formation concerning projects similar to the one proposed, in which the WAM membrane was installed, including location and person to be contacted.

B. *Inspections:* Third party inspection of the installation is recommended. A list of SPFD/SPI or CUFCA inspectors is available.

1.2 Submittals

A. The WAM membrane should be included in the data sheets or letter of certification that the product comply with the materials specified.

B. The WAM membrane should be included in the shop drawings on construction details.

C. Safety and handling instructions for storage, handling, and use of the WAM materials to include appropriate Materials Safety Data Sheets (MSDS).

1.3 Materials Storage

A. Materials shall be adequately protected and stored at all times in a dry space, properly ventilated and protected from the elements. In winter conditions, air temperature should not fall below 10°C.

1.4 Related Work

1. Concrete: Section 03300, Cast-in-Place Concrete
2. Structural Steel and Steel Joists: Section 05120
3. Metal Decking: Section 05300, Steel Deck
4. Masonry: Section 04200
5. Firestopping: Section 07270
6. Insulation: Section 07210
7. Drywall: Section 09250, Gypsum Wallboard
8. Caulking: Section 07900, Sealants

1.5 Warranty

1. Warrant work of this section against defects or deficiencies for a period of two years. Work is certified as substantially performed in accordance with General Conditions of the Contract.
2. Promptly correct, at own expense, defects or deficiencies, which become apparent within warranty period.

2. MATERIALS

1. A high performance, either self-adhesive (SA) or thermofusible grade (TG)[81] air barrier membranes with the following minimum performance characteristics:
   - Tensile strength for TG membrane
     —Machine direction (md) 5.5 N/mm (275 N/5 cm)
     —Xd 5 N/mm (250 N/5 cm)
   - Breaking strength for SA membrane
     —Machine direction (md) 3 N/mm (150 N/5 cm)
     —Xd 3 N/mm (150 N/5 cm)
   - Low temperature flexibility −30°C (−22°F)
   - Peel Adhesion (7 days, temperature 5 to 45°C) concrete, metal or itself
     —Minimum 1.5 N/mm (75 N/5 cm)

2. The selected WAM membrane shall not be provided with a polyethylene on the bonding surface, as the SPF adheres poorly to the polyethylene surface. A brief flame-treatment of the exterior, polyethylene separation sheet may be necessary before applying the SPF.

3. EXECUTION

3.1 Surface Preparation

1. Examine areas before applying the WAM membrane to ensure that the configuration, surfaces, dimensions, and environment are suitable for membrane installation; and that execution of installation and its performance will not be adversely affected.
2. Surfaces to receive the WAM membrane shall be dry, frost-free and not coated with releasing agents or other coating affecting bond. Concrete surface must be at least ten days old. Report in writing defects in work including adverse conditions. Do not work until such conditions are rectified. Application of work of this section shall be deemed acceptance of existing work and existing conditions.
3. Prepare surface by brushing, scrubbing, scraping, or grinding to remove loose mortar, dust, oil, grease, oxidation, mill scale, and other materials which will affect adhesion and integrity of foam insulation. Clean metal

---

[81]Typically SBS-modified bitumen membrane reinforced with non-woven fiberglass, polyester, glass scrim, or high density polyethylene film. Examples are: Blueskin SA and Blueskin TG by Bakor Inc., Sopraseal Stick and Sopraseal 180HD by Soprema Inc., Bituthane 5000 by W.R. Grace & Co., or equivalent.

surfaces to remove release agents or other noncompatible coatings, using clean cellulose sponges or rags soaked in solvent compatible with the WAM membrane and SPF. Ensure surfaces are dry before proceeding.

4. All major cracks, openings, and gaps must be filled prior to installation of the WAM membrane.
5. If a width of the crack or opening is in excess of 6 mm (1/4 in), fill it with the BSF type of SPF, making allowance for post-expansion of the foam. Use manufacturer recommended dispensing gun.
6. Do not allow BSF to cover or spot surfaces beyond the opening to be filled. Use masking materials if necessary.
7. Remove BSF droppings etc., beyond the opening, and remove the masking tape, immediately after BSF has cured to hard surface film. Clean and make good surfaces soiled or damaged by work with BSF type SPF.

Note to the Specifier: Show on drawings or otherwise specify locations of WAM membranes.

3.2 Installation

1. Apply WAM membrane in strict accordance with manufacturer's written instructions, specifications or recommendations. If primer is needed (e.g., concrete, masonry), apply primer to surface at the rate 0.25 liters/m^2 and allow to dry.
2. A minimum 50 mm overlap is required for joining the membrane segments. A minimum distance of 125 mm (5 inch) from the crack, joint or edge of the wall is required for establishing a sufficient connection between various component parts of the building envelope.
3. To minimize a number of overlaps between the membrane strips which span the opening, a strip of minimum 300 mm wide and 3 m long is recommended for areas where membrane is unsupported (around windows, edge beams, etc.).
4. Where installation cannot be carried out using a torch, use a self-adhesive (SA) membrane, using a priming agent, as recommended by the manufacturer, prior to membrane installation. When SA membrane joints the TG, the SA should overlap on the top of the other, and joint be sealed with mastic.
5. Install the SPF as soon as possible following the inspection of the membrane by a qualified personnel.
6. All WAM membranes must be protected from excessively high temperatures.

Guide and Selection Criteria for SPF Used in Building Envelopes (Roofs and Walls)

1. GENERAL CONSIDERATIONS

Performance of a spray applied polyurethane foam roofing system can be affected by all the component parts of a roof structure, as well as the service and climatic conditions inside and outside the structure. Proper structural design, specification review, contractor and material selection, coupled with the compatibility and positioning of the various components of a roof structure, are necessary to produce a successful roofing system.

Consult with material suppliers and the successful contractor to receive written confirmation of their agreement to all aspects of the roofing system, including, but not limited to, material selection, drainage, expansion joints, load design, flashing details, deck preparation, etc.

In residential and office spaces, under normal heating and cooling conditions, vapor retarder, traditionally placed on the warm side may not be required and moisture control is achieved with a "flow-through" design principle (see Chapter 8).

2. PROCEDURES AND CONSIDERATIONS FOR ROOFING APPLICATIONS

SPF can be applied to most surfaces. However, the following general practices must be observed on all decks prior to receiving SPF.

2.1 General Surface/Deck Preparation Procedures

- The roof deck shall be securely fastened to the building structure and conform to proper load limits of good engineering practices. Attention should be paid to the deflection, including but not limited to, foot traffic, mechanical equipment utilization, live and dead loads as well as resistance to dynamic forces of wind uplift.

309

- When a primer and/or a vapor retarder is specified, there must be adequate adhesion between all components of the system to secure the entire system against wind uplift and structural movement.
- Prior to application of primer, vapor retarder, or SPF, the deck shall be properly cured, dry, and free of any contaminants that may interfere with proper adhesion of any of these respective components.
- Deck contaminants, depending on their severity and quantity, may be removed by use of air pressure, vacuum equipment, hand power broom, chemical solvents, sandblasting, manual scraping, etc.

2.2 Wood Surfaces/Decks

- A pretreatment with a primer is necessary to achieve maximum adhesion of the SPF to a newly constructed wood deck.
- Joints in excess of 1/4" in width shall be filled and sealed prior to the application of the respective primer, vapor retarder (if necessary), or SPF.

2.3 Metal Surfaces/Decks

- A pitch of 1/4" in 12" or more is recommended.
- A structural metal deck should not be lighter than 22 gauge.
- Sloped metal roof panels should not be lighter than 29 gauge.
- All joints should be correctly lapped, sealed, and fastened.
- Underlay board, if specified for smoother application of SPF should be of sufficient width and thickness to span or fill flutes. Fastening shall be in accordance with applicable code requirements.

2.4 Concrete Surfaces/Decks

- In all cases, concrete should be free of laitance and chemical release agents.
- Priming is required on concrete surfaces, and it is recommended that due to the water of hydration that is present, poured concrete decks be permitted to cure for twenty-eight (28) days prior to the application of SPF.
- All joints should be filled and/or taped.
- SPF can be directly applied to the lightweight or insulating concrete only when tests have been made to determine that adequate adhesion can be obtained. Normally, a mechanically fastened underlay board is used.

2.5 Selection of a Protective Coating

When SPF is applied as an integral part of the roofing system, it must be given a protective layer. Typically, the required protection is attained through application of an elastomeric liquid applied coating system, following the manufacturer's recommendations, though it may also be protected by other means.

The average and the minimum of the dry film thickness (DFT) of the protective coating shall comply with the coating manufacturer's specification. The properties of the cured protective coating shall meet the minimum design characteristics of the generic type specified. The protective coating shall be specifically manufactured for the water and UV protection of SPF as used in roofing applications.

When selecting the coating materials, the specifier should consider the following:

1. Material physical characteristics
   - Chemical resistance
   - Water vapor permeance
   - Tensile and elongation properties
   - Retention of physical properties upon aging
   - UV resistance

2. Environment and performance characteristics
   - Environment in which to be used (abuse, hail resistance, etc.)
   - Life expectancy
   - Ease of maintenance
   - History of similar applications or laboratory data relating to the application in question
   - Adhesion to the SPF
   - Combustibility characteristics, individually and in combination with the selected SPF systems
   - Ability to withstand foot traffic
   - Aesthetic qualities

2.6 Maintenance Procedures

It is strongly recommended that maintenance procedures, including annual inspections, be established with your selected contractor for any roofing system to yield its full value. The purchaser/specifier should contact the respective manufacturer/supplier and contractor for their recommended maintenance procedure.

3. SELECTION OF THE SPF SYSTEM

A purchaser is buying a finished roofing insulation system. The contractor is fabricating the product on site in accordance with manufacturer's instructions. A range of SPF systems are available in various densities, exhibiting different temperature limitations, and combustibility characteristics, provided with a wide range of economical installations.

A purchaser should understand that most published data is obtained on laboratory samples. The conditions of foam installation such as thickness and number of passes, substrate and ambient temperatures, have a pronounced effect on all properties; the purchaser should therefore insist on appropriate field quality assurance.

From a fire safety standpoint, SPF can be used safely. It is important, however, that all persons associated with the design, fabrication, storage, and installation understand the materials and environments involved. SPF is a combustible product. Care must be taken to ensure that the foam is not exposed to a heat source or an open flame.

4. CHECK-LISTS

Two checklists (Tables A1 and A2) provided in this appendix neither summarize nor replace the information collected in Chapters 8, 9, and 10. For instance, typical considerations during a selection of SPF roof covered with an elastomeric coating (SPFD, 1989, 1990) are listed in Table 41 in Chapter 9. Yet the checklist is a practical tool that enhances the holistic approach to the design.

Table A1. Typical considerations during a selection of SPF roof/elastomeric coating.

1. Building type and occupancy: mid- or high-rise residential, office, warehouse or industrial
2. Accessibility of roof area: height of building, roof sculters, ladders, hatches and surrounding structures
3. Weight of the roofing system, structural, utility and snow loads
4. Climate of the building surroundings: mean monthly temperatures and daily variations, potential for high winds and hail, proximity to chemical emissions
5. Traffic on the roof surface: foot and rolling traffic (e.g., window washing equipment)
6. Penetrations and roof-mounted equipment: vents and plumbing stacks, HVAC, fans, filter units, skylights, cooling tower, pipes and conduits and chimneys
7. Roof configuration and details: roof slope, locations of drains and overflows, flashing details, and expansion joints
8. Review prior to the SPF roof installation:
   • compatibility with adjacent construction, siding, parapet walls, flashing materials
   • type and performance of the substrate: strength and dimensional stability
   • adhesion or mechanical fasteners bonding the substrate to the deck (wind uplift resistance)
9. Required level of physical characteristics of coating materials: *(continued)*

Table A1. (continued).

- water vapor permeance, chemical resistance, adhesion to SPF, UV resistance
- tensile and elongation properties and their retention during the service life
10. Restrains during the SPF installation:
    - containment, penetrations, faults, voids and defects of the substrate
    - substrate surface: contamination (to be cleaned or sand blasted), moisture content
    - conditions during the SPF installation: wind speed, temperature and relative humidity
11. Requirements of SPF surface: color and texture of the finished surface, surface IR-reflectivity
12. Required design life and selection of the maintenance and repair model
13. Service and maintenance considerations:
    - modifications (see Chapter 4) and repairs can be made by in-house maintenance personnel
    - re-coating (see Chapter 10) to bring the roofing system to its original performance level
14. Life cycle economics: initial cost (installation), service and maintenance cost

---

Table A2. Typical considerations during a selection of SPF for walls.

---

1. Climate of the building surroundings: mean monthly temperatures, and daily variations
2. Building type and occupancy: mid- or high-rise residential, office, warehouse, or industrial
3. Building height and exposure to wind: maximum wind loads and stack pressures
4. Moisture considerations related to external or internal placement of the SPF barrier within the cross-section of the wall
5. Review prior to the SPF installation:
    - continuity of air, moisture, and thermal barriers
    - construction details: window/wall interface, penetrations, floor and wall junctions
    - compatibility with adjacent construction
    - type and performance of the substrate: strength and dimensional stability
6. Restrains during the SPF installation:
    - containment, penetrations, faults, voids and defects of the substrate
    - substrate surface: contamination (to be cleaned or sand blasted), moisture content
    - conditions during the SPF installation: wind speed, temperature, and relative humidity
7. Required design life
8. Service and maintenance considerations
9. Life cycle economics: initial cost (installation), service, and maintenance cost

---

The Influence of Environmental Factors on Thermal Resistance of SPF

The influence of environmental factors has yet not been discussed in this book. This issue is particularly important in reformulating the polymeric system to optimize their performance with a specific blowing agent or a combination of blowing agents.

An overview, given in a paper by Ascough, Bomberg and Kumaran (1991) reviewed performance of one generic SPF system, whose composition was presented in Table 8, manufactured with CFC-11, HCFC 123, HCFC-141b and exposed to different environmental conditions.

TEMPERATURE AND BA PRESSURE AT
THE BEGINNING OF CONDENSATION (BC)

Figure A9 presents temperature dependence of thermal resistivity (inverse of thermal conductivity) of a typical polyurethane foam. A computer-controlled Heat Flow Meter (HFM) apparatus was used to simultaneously shift temperature on both sides of the specimen, in a stepwise fashion. To increase the precision of determination of the BC point, test specimen was 10 mm thick and temperature difference applied to the specimen was also smaller than that postulated in ASTM C518 standard.

To ensure the validity of experimental comparisons, one must establish the magnitude of material variability and its effect on the measured property (for instance, variation in thermal resistivity caused by the use of different specimens). Figure A10 shows measurements performed on two additional specimens that were tested initially (curves 1 and 2), and after 30 days two different exposures (curves 3 and 5). These results were compared with those obtained on other test specimens: curve 4 that was established after 30 days of room exposure and curve 6 determined after 30 days in dry cube exposure. The comparison illustrates a negligible effect of specimen selection in relation to the differences caused by various environmental exposures.

In short, as shown in Figure A10, precision of testing thermal resistivity as a function of temperature is sufficiently high.

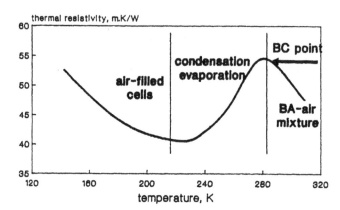

Figure A9. Temperature dependence of thermal resistivity (inverse of thermal conductivity) of a typical SPF; from Ascough et al. (1991).

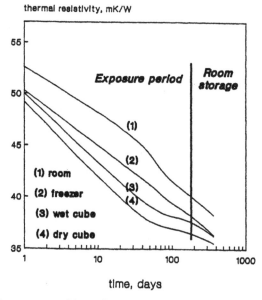

Figure A10. Measurements of thermal resistivity versus time performed on different specimens. Materials and environmental conditions are described later in the text; from Ascough et al. (1991).

Table A3. Shift in BC for HCFC-141b blown foams over 9–12 months period.

Encapsulated after Days	Apparent Shift in BC Point, °C	Long-Term BC Value, °C
11	11.5	5.0
16	14.0	4.5
25	12.5	4.8

Table A3 shows results of measurements on this SPF system manufactured with HCFC-141b and performed over 9- to 12-month period. The long-term BC value established at the end of this period resulted in partial pressure of the BA about 0.33 atmosphere.

EFFECT OF ENVIRONMENTAL FACTORS ON AGING OF SPRAY POLYURETHANE FOAM

All SPF types had identical polymeric system as one presented in Table 8 but used different BA combinations, namely:

- CFC 11 alone, a generic SPF to which others were compared
- HCFC 123
- 60/40 blend of HCFC 123/HCFC-141b plus 0.5% water
- a proprietary blend of HCFC 123/HCFC-141b

The slicing, encapsulation and testing procedures (Kumaran et al., 1989; Bomberg et al., 1990) were used to measure thermal resistance after exposure to various environmental conditions for selected intervals.

These environmental conditions were:

- isothermal conditions: temperature equal to $21 \pm 1°C$, and relative humidity of $50 \pm 10\%$, termed room exposure
- thermal gradient without moisture: one side of the specimen maintained at $70 \pm 2°C$ (dry), the opposite at room conditions, termed dry cube exposure
- thermal gradient with moisture flow: one side of the specimen maintained at $70 \pm 2°C$ and relative humidity above 95% RH, the opposite at room conditions, termed wet cube exposure
- thermal gradient with freeze-thaw: one side of the specimen subjected to freeze-thaw cycle $-20°C$ for six hours and then slowly increasing temperature up to $5–10°C$ during the next 6 hours, the opposite at room conditions, termed freeze-thaw exposure

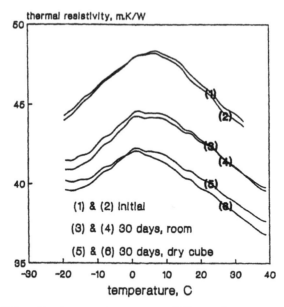

Figure A11. Effect of environmental exposure on SPF manufactured with 60/40 blend of HCFC 123/HCFC-141b plus 0.5% water; from Ascough et al. (1991).

The specimens were encapsulated with epoxy on all surfaces except for one that was exposed to the prescribed exposure conditions. Periodically, thermal measurements were performed using two specimens exposed to identical conditions and placed one over the other. Keeping the open surfaces together, the edges of the specimens were sealed. After completion of the thermal measurements the edges were opened, specimens were separated, and placed back for continuing their exposure. After 6-month period of exposure, all specimens were placed in a room for further 9 to 12- month.

Figure A11 shows the effect of environmental exposure on SPF manufactured with 60/40 blend of HCFC 123/HCFC-141b plus 0.5% water.

Figure A10 shows changes in thermal resistivity measured on 25 mm thick SPF specimens manufactured with 60/40 blend of HCFC 123/HCFC-141b plus 0.5% water exposed to four different environmental conditions. There appears to be no difference in the rate of aging of all specimens. There is, however, slightly higher R-value of the specimen in the room conditions (1). The differences between each specimen tested after 6-month exposure and another 6 months in the room storage, i.e., after one year appear to be almost the same as in the beginning of the exposure. In other words, the difference between specimens appears to be much more affected by the specimen itself rather than the exposure conditions.

Similar tests were performed on SPF using other three BA. These tests indicated that SPF manufactured with CFC 11 had somewhat different behavior than those of SPF manufactured with various HCFC combinations. The cause of this difference is not clear, though could be ascribed to an interaction between different factors affecting BA solubility.

As shown above, environmental conditions have a marked effect on the thermal performance of SPF irrespective of the blowing agent used. The dependence of the aging process on the environment to which the foam is exposed was demonstrated by measuring their thermal resistivity at selected intervals.

A number of environmental and material factors influencing the apparent thermal resistance of foams were identified (Hilado, 1967; SPI, 1988; Kumaran et al. 1989; Bomberg at al. 1990). The environmental factors are: mean temperature and temperature gradient, moisture content and rate of moisture migration through the material. The material factors are: composition of cell gas gases and partial pressure of each of them. In turn, these material factors depend on the initial concentration of the blowing agents, permeability of the polymeric matrix to oxygen, nitrogen and blowing agents and the mass of blowing agent stored within the polymer matrix.

All these processes affect thermal performance but each of them occurs at different rate (SPI, 1988). Therefore, during a certain period after any change in conditions on the material surface (e.g., change of temperature or temperature gradient) gaseous, adsorbed and absorbed fractions of the cell gas components may not be in equilibrium. It is important to differentiate between the dynamic and stabilized values of thermal resistance (the latter occurs when the equilibrium is established). Figure A11 showed the dynamic value during the 180-days of exposure and the stabilized value after the next 180-day storage in the laboratory conditions.

It is difficult to correlate the dynamic thermal performance measured under strong environmental effect such as exposure to elevated temperature. This fact must be conveyed to all these standardization committees, who either maintain, or try to reintroduce, determination of "aged" (long-term) thermal resistance based on exposure to elevated temperature.

In closing, the authors presented a number of new techniques developed during the CPIA/NRC research program, namely
- the BC method of non-destructive measurement of BA partial pressure
- the normalized aging curve determined on thin foam layers, a physical characteristic of the foam product
- dry cube, wet cube and freeze-thaw exposures to examine effects of selected environmental factors
- the scaling factors or distributed parameters continuum (DIPAC) model to

predict long-term thermal resistance of any foam/blowing agent combination

These techniques permit qualifying long-term thermal performance of current SPF systems as well as any other polymeric foam irrespective of the blowing agent selected.

AIA. 1969. *Architects' Handbook of Professional Practice,* Chapter 11, "Project Procedures," The American Institute of Architects, 1735 New York Avenue, N.M., Washington, D.C. 20006.

Allen, D. E. and M. T. Bomberg. 1997. "Limit States Design for Durability," *Proceedings 7th Conf. on Building Science and Technology,* March 21, 1997, Toronto, ON.

Andersen, J.1993. "Spray Polyurethane Improves Racking Strength of Cavity Walls," *J. Thermal Insul. and Bldg. Envs.,* 16:103–105.

Ascough, M. R., M. T. Bomberg and M. K. Kumaran. 1991. "Effect of Time and Temperature on R-Value of Rigid Polyurethane Foam Insulation Manufactured with Alternative Blowing Agents," *Polyurethane World Congress,* pp. 399–407.

ASHRAE Handbook. 1985. Fundamentals, SI Edition, Chapter 23. "Design Heat Transmission Coefficients."

ASTM Standard D 5116. 1990. "Standard Guide for Determination of VOC Emissions in Environmental Chambers from Materials and Products."

ASTM Standards. Chapter 12.2. "ASTM Standards Used by the SPF Industry."

Alumbaugh, R. L. and J. R. Keeton. 1977. Investigation of Spray-Applied Polyurethane Foam Roofing Systems I, Naval Civil Engineering Laboratory, Port Hueneme, N-1496, pp. 1–65.

Alumbaugh, R.L, J. R. Keeton and E. F. Humm.1983. Experimental Polyurethane Foam (PUF) Roofing Systems II, Naval Civil Engineering Laboratory, Port Hueneme, N-1656, pp. 1–51.

Alumbaugh, R.L, S. R. Conklin and D. A. Zarate. 1984. Preliminary Guidelines for Maintenance of Polyurethane Foam (PUF) Roofing Systems, Naval Civil Engineering Laboratory, Port Hueneme, N-1691, pp. 1–42.

Alumbaugh, R. L., E. F. Humm and J. R. Keeton. 1990. Investigation of Spray-Applied Polyurethane Foam Roofing Systems II, Naval Civil Engineering Laboratory, Port Hueneme, N-1815, pp. 1–72.

Alumbaugh, R.L. 1993. *Field Performance of Spray Applied Thermal Insulating Foams,* The Soc. of the Plastics Industry, Inc., Polyurethane Foam Contractors Div., Washington, D.C., pp. 29–77.

Baker, M. C. 1980. *Roofs, Design, Application and Maintenance,* Polyscience Publications Inc.

Ball, G. L. III, N. Schwartz and J. S. Long. 1960. "The Sound Absorption Properties of Urethane Foams," *Official Digest,* pp. 817–831.

Bankvall, C. G. 1986. "Air Movements and the Thermal Performance of the Building," *ASTM STP 922,* American Soc. for Testing and Materials, pp. 124–131.

Bankvall, C. G. 1986a. "Thermal Performance of the Building Envelope as Influenced by Workmanship," *ASTM STP 922,* American Soc. for Testing and Materials, pp. 679–684.

Baskaran, B. A. and O. Dutt. 1995. "Evaluation of Roof Fasteners under Dynamic Loading," *Ninth International Wind Engineering Conference,* January 9–13.

Baumann, G. F. 1982. "R-Value of Rigid Polyurethane Foam: An SPI Research Project," *Journal of Thermal Insulation,* 6:39–46.

Bearg, D. W. 1993. *Indoor Air Quality and HVAC Systems,* Lewis Publishers, Boca Raton, FL, p.74

Becker, R. 1985. "A Method for the Generation of Weighing Factors for Performance Evaluation Systems," *Building and Environment,* 20(4):195–200.

Bement, L. B. and R. C. Parker. 1994. "Adhesion of Weather Proofing Coatings to HCFC-141b Blown Spray Foam," *J. Thermal Insulation and Bldg. Envelopes,* 18:70–80.

Bergin, M. H., D. Y. H. Pui, T. H. Kuehn and W. T. Fay. 1989. "Laboratory and Field Measurements of Fractional Efficiency of Industrial Dust Collectors," *ASHRAE Transactions,* 9(2):102–112.

Blach, K. 1972. "How Can Performance Be Specified?" Danish Building Research Institute, Stencil.

Blach, K. and G. Christensen. 1976. "The Performance Concept," *Building Research and Practice,* May/June, pp. 152–166.

Bloomberg, T. and J. Claesson. 1993. "Metal Thermal Bridges in Thermal Insulation." Lund U., presented at *CIB W40 Meeting* at Budapest.

Bomberg, M. and C. J. Shirtliffe. 1978. "Influence of Moisture and Moisture Gradients on Heat Transfer through Porous Building Materials," *ASTM Special Technical Publication 660,* pp. 211–233.

Bomberg, M. T. 1980. "Problems in Predicting the Thermal Properties of Faced Polyurethane Foams," *Thermal Insulation Performance, ASTM, STP 718,* pp. 412–428.

Bomberg, M. 1982. "Development of Thermal Insulation Performance Test Methods," *ASTM Standardization News,* Dec. , pp. 26–32.

Bomberg, M. T. 1983. "Laboratory Methods for Determining Moisture Absorption of Thermal Insulations, I: Review," *J. of Thermal Insulation,* 6:232–249.

Bomberg, M. T. and M. K. Kumaran. 1985. "A Test Method to Determine Air Flow Resistance of Exterior Membranes and Sheathings," Building Research Note 227, National Research Council of Canada.

Bomberg, M. and N. V. Schwartz. 1986. "Building Envelope with Sprayed Cellular Insulation, a Program for Developing Evaluation Procedures," *J. Thermal Insulation,* 9:83–90.

Bomberg, M. T. 1988. "A Model of Aging of Gas-Filled Cellular Plastics," *Journal of Cellular Plastics,* 24(4):327–347.

Bomberg, M. T. and M. K. Kumaran. 1989. "Report on Sprayed Polyurethane Foam with Alternative Blowing Agents," *CFCs and the Polyurethane Industry: Volume 2, A Compilation of Technical Publications,* SPI, p. 112–128.

Bomberg M., 1974. "Moisture Flow through Porous Building Materials," Lund Inst. of Techn., Lund, Sweden, 52:1–88.

Bomberg, M. T. 1990. "Scaling Factors in Aging of Gas-Filled Cellular Plastics," *Journal of Thermal Insulation,* 13:149–159.

Bomberg, M. T. and D. A. Brandreth. 1990. "Evaluation of Long-Term Thermal Resistance of Gas-Filled Foams: Insulation Materials, Testing, and Applications," *ASTM STP 1030,* pp. 156–173.

Bomberg, M. T., M. K. Kumaran, M. R. Ascough and R. G. Sylvester. 1991. "Effect of Time and Temperature on *R*-Value of Rigid Polyurethane Foams," *Journal of Thermal Insulation,* 14:343–358.

Bomberg, M. 1993. *Factors Affecting the Field Performance of Spray Applied Thermal Insulating Foams,* The Society of the Plastics Industry, Inc., Polyurethane Foam Contractors Division, Washington, D.C., pp. 29–77.

Bomberg, M. and W. C. Brown. 1993. "Building Envelope and Environmental Control," *J. Thermal Insul. and Bldg. Envs.,* 16: 306–311 and 17:5–12.

Bomberg, M. T. and M. K. Kumaran. 1994. "Laboratory and Roofing Exposures of Cellular Plastic Insulation to Verify a Model of Aging," *3rd Symp. on Roofing Res. and Standards Development, ASTM STP 1224,* T. J. Wallace and W. J. Rossiter, Jr., eds., Philadelphia, ASTM, pp. 151–167.

Bomberg, M. and C. J. Shirtliffe. 1994. "A Conceptual System of Moisture Performance Analysis," Chapter 26, *ASTM Manual Series MNL 18, Manual on Moisture Control in Buildings,* pp. 453–461.

Bomberg, M. T. and M. K. Kumaran. 1995. "Use of Distributed Parameter Continuum (DIPAC) Model for Estimating the Long-Term Thermal Performance of Insulating Foams," *Proceedings of the Thermal Performance of the Exterior Envelopes of Buildings VI,* Dec 4–8, Clearwater Beach, FL, ASHRAE.

Bomberg, M. T. and D. E. Allen. 1996. "Use of Generalized Limit States Design of Building Envelope for Durability," *J. of Thermal Insul. and Bldg Envs.,* 20:18–39.

Booth, L. D. and W. M. Lee. 1986. "Effects of Polymer Structure on *k*-Factor Aging of Rigid Polyurethane Foam," *J. Cellular Plastics,* 20:26 or *SPI Conference Polyurethane 84,* p. 268.

Booth, J. R. 1991. "Some Factors Affecting the Long-Term Thermal Insulating Performance of Extruded Polystyrene Foams," *ASTM STP 1116,* pp.197–213.

Booth, J. R. and T. S. Holstein. 1993. "Determination of Effective Diffusion Coefficients of Nitrogen in Extruded Polystyrene Foam by Gravimetric Sorption," *J. of Thermal Insul. and Bldg. Envs.,* 16:246–262.

Brand, R. 1990. *Architectural Details for Insulated Buildings,* Van Nostrand Reinhold, NY.

Brandreth, D. A. and H. G. Ingersoll. 1980. "Accelarated Aging of Rigid Polyurethane Foam," *European J. of Cellular Plastics,* pp.134–143.

Brandreth, D. A. 1981. "Factors Influencing the Aging of Rigid Polyurethane Foam," *J. of Thermal Insulation,* 5:31–39.

Braun, R., J. Hansen and A. Woods. 1995. "Urethane Foams for Air Leakage Control," *Home Energy,* pp. 25–28.

Brehm, T. M. and L. R. Glicksman. 1989. "A New Sorption Technique for Rapid Measurement of Gas-Diffusion and Solubility in Closed-Cell Foam Insulation," *Polyurethane 89*, SPI USA, pp. 547–552.

Brown, W. C. 1986. "Heat-Transmission Tests on Sheet Steel Walls," *ASHRAE Transactions*, 92(2B):554–566.

Brown, W. C. and N. V. Schwartz. 1987. "The Thermal and Air Leakage Performance of Residential Walls," presented to *SPI Conference in Huntsville*, Ont, IRC paper No. 1527.

Brown, W. C. and G. F. Poirier. 1988. "Testing of Air Barrier Systems for Wood Frame Walls," Report to CMHC, NCR report CR 5505.1:81

Brown, W. C. and D. G. Stephenson. 1993. "A Guarded Hot Box Procedure for Determining the Dynamic Response of Full-Scale Wall Specimens," *ASHRAE Transactions: Research*, I:632–642; II:643–660.

Brown, W. C., M. T. Bomberg, J. M. Ullett and J. Rassmussen. 1993. "Measured Thermal Resistance of Frame Walls with Defects in the Installation of Mineral Fibre Insulation," *J. of Thermal. Insul. and Bldg. Envs.*, 16:318–339.

Brown, W. C., P. Adams, T. Tonyan and J. Ullet. 1997. "Water Management in Exterior Wall Claddings," *J. of Thermal. Insul. and Bldg. Envs.*, 21:23–44.

Burns, S. B., S. N. Singh and I. D. Bowers. 1998, "The Influence of Temperature on Compressive Properties and Dimensional Stability of Rigid Polyurethane Foams," *J. Cellular Plastics*, 134:18–38.

CAN/CGSB 12.8 M90 Insulating Glass Units.

Cartmell, M. J. and R. K. Brown. "Polyurea Foam for Retrofit Insulation—An Update with Case Studies," *J. Thermal Insulation*, 6:250–262.

Chand, I., V. K. Sharma and P. K. Bhargava. 1995. "Effect of Neighboring Buildings on Mean Wind Pressure Distribution on a Flat Roof," *Architectural Science Review*, 38: 29–36.

Christian, J. E., G. E. Courville, R. S. Graves, R. L. Linkous, D. L. McElroy, F. J. Weaver and D. W. Yarbrough. 1991. "Thermal Measurement of in-situ and Thin Specimen Aging of Experimental Polyisocyanurate Roof Insulation Foamed with Alternative Blowing Agents," *ASTM STP 1116*, pp. 142–166.

Conklin, S. R., D. A. Zarate and R. L. Alumbaugh. 1986. Experimental Polyurethane Foam (PUF) Roofing Systems III, Naval Civil Engineering Laboratory, Port Hueneme, N-1742, pp. 1–42.

Coultrap, K. and R. L. Alumbaugh. 1987. Users Guide for Foam Roofing, U.S. Navy, Naval Civil Engineering Laboratory, UG-0011, pp. 1–77.

Cullen, W. C. and W. J. Rossiter. 1973. Guidelines for Selection of and Use of Foam Polyurethane Roofing Systems, National Bureau of Standards (NBS) Technical Note 778, pp. 1–23.

Cullen, W. C. and T. Sneck. 1980. "A Final Report of RILEM Technical Committee 27-EVS, the Evaluation of External Vertical Surface of Buildings," *Bulletin RILEM, Materiaux et Construction*, 13(76).

Dechow, F. J. and K. A. Epstein. 1978. "Laboratory and Field Investigations of Moisture Absorption and Its Effect on Thermal Performance of Various Insulations," *ASTM, STP 660*, pp. 234–260.

Desjarles A. et al., 1995.

Detail-Types. 1997. Conception pour l'Envelope du Batiment (Details and Solutions for Building Envelope), Airmetic 0223, Demilec.

Di Lenardo, B., W. C. Brown, W. A. Dagliesh, G. F. Poirier and M. K. Kumaran. 1995. "Air Barrier Systems for Exterior Walls of Low-Rise Buildings," *CCMC Technical Guide Master Format 07195*, Institute for Research in Construction, NRC Canada.

Di Lenardo, B. and M. T. Bomberg. 1995. "The Reference Period for Predicting Long-Term Thermal Performance of Thermal Insulating Foams," *J. Cellular Plastics*, 31: 356–374.

Dillon, P. W. 1978. Private Communication, Chemicals and Plastics, Research and Development Department, Tarrytown, NY, May.

Edgecombe, F. H. 1989. "Progress in Evaluating Long-Term Thermal Resistance of Cellular Plastics," *Polyurethanes 89, SPI Conference*, pp. 676–683.

Eliott, R. W. "Corrosivity of Polyurethane Foam toward Metals," *J. Thermal Insulation*, 10:142–146.

European Passive Solar Handbook. 1986. Commission of the European Communities, Directorate General XII for Science, Research and Development, Solar Energy Division, Brussels.

Fang, J. B., R. A. Grot, K. W. Childs and G. E. Courville. 1984. "Heat Loss from Thermal Bridges," *Building Research and Practice, CIB*, 12(6):346–352.

Flynn, J. E. and A. W. Segil. 1970. *Architectural Interior Systems, Lighting, Air Conditioning, Acoustics*, Van Nostrand Reinhold, NY.

Forgues, A. O. 1983. "Laboratory Methods for Determining Moisture Absorption of Thermal Insulations. II: Comparison of Three Water Absorption Test Methods with Field Performance Data," *Journal of Thermal Insulation*, 7:128–137.

Franklin Associates. 1991. Comparative Energy Evaluation of Plastic Products and Their Alternatives for the Building and Construction and Transportation Industries, Report for the Society of Plastics Industry Inc.

Fraser, H. W. and J. C. Jofriet. 1991. "Thermal Behaviour of Polyurethane Foam Insulation when Subjected to a Thermally Induced Vapour Pressure Gradient and Freeze/Thaw Cycling," *Canadian Society of Agricultural Engineering Meeting*, July 29–31, paper 91–220, pp. 1–28.

Freon. 1985. Properties of Rigid Polyurethane Foams, Freon Product Information, BA-13, DuPont.

Gaarenstroom, P. D., J. B. Lets and A. M. Harrison. 1989. "Distribution of CFC-11 in Aged Urethane and Isocyanurate CLBS Foam," *Polyurethanes 89*, Union Carbide Corp., pp. 163–168.

Garrett, K. W. 1979. *An Assessment of the Calculation Methods to Determine the Thermal Performance of Slotted Building Blocks*, Build. Services Eng. Res. and Technology, pp. 24–30.

Gertis, K. 1982. Building Physics: Trends and Future Tasks for Civil Engineers and Architects, Int. Assoc. for Bridge and Structural Eng., S-21/82.

Gibson, L. J. and M. F. Ashby. 1988. *Cellular Solids, Structure and Properties*, Pergamon Press, pp. 1–344.

Gibson, J. 1993. "Pressurization and Air Control," *Construction Canada*, 3:14–16.

Glicksman, L. R., M. Burke, A. Marge and M. Mozgowiec. 1991. "A Review of

Techniques for Improved Foam Conductivity: Reducing Radiation Heat Transfer, Limiting Aging and Inclusion of Vacuum Elements, Thermal Insulation Performance," *ASTM STP 1116*, pp. 237–259.

Greason, D. M. 1983. "Calculated versus Measured Thermal Resistances of Simulated Building Walls Incorporating Airspaces," *ASHRAE Transactions,*89(1), preprint AC-83-12, No. 3.

Grodin. 1993. Damage Functions for Service Life Prediction of Zinc and Steel Components, Internal Report No. 647, Institute for Research in Construction NRC.

Hallee, N. D. and J. Hunter. Potato Storage Design and Management, Bulletin 656, U. of Maine at Orno.

Harrison, H. W. 1983. "Practical Problems in Estimating Service Life of Components and Elements in Housing," *To Build and Take Care of What We Have Built with Limited Resources, Building Technology Design and Production, Vol. 2,* pp. 81–90, CIB; see also R.B. Bonshor and H.W. Harrison, 1982, *Quality in Traditional Housing, Vols. 1–3,* HMSO, London.

Hedlin, C. P. 1977. "Moisture Gains by Foam Plastics Roof Insulations under Controlled Temperature Gradients," *J. Cellular Plastics,* 13:313–319.

Hens, H. 1992. "IEA Annex 14, Condensation and Energy," *J. Thermal Insulation,* 15:261–273.

Hilado, C. J. 1967. "Effect of Accelerated and Environmental Aging on Rigid Polyurethane Foam," *J. Cellular Plastics,* 3:161–167.

Hollingworth, M., Jr. "Experimental Determination of the Thermal Resistance of Reflective Insulations," *ASHRAE Transactions,* 89(1), preprint AC-83-12 No. 2.

Hutcheon, N. B. 1953. Fundamental Considerations in the Design of Exterior Walls in Buildings, National Research Council of Canada, Div. Building Research, Tech. Report No. 13; see also "Philosophy of Design and Testing of Joints," *CIB Symposium on Joints,* Oslo 1967.

Hutcheon, N. B. 1971. "The Utility of Building Science," Invited Address at the Central Building Research Institute, India, October; see also *Symposium on Panelized Building Systems,* June 2–3, 1970 at Sir George Williams University, Montreal, Canada.

ICI Polyurethanes. 1995. Physical Testing Method for Dimensional Stability, Method AG05, Procedure B.

Isberg, J. 1977. "Sandwich Panel Containing Polyurethane Foam, Thermal and Moisture Investigations, Test House in Landsbro," (in Swedish), Chalmers University, Gothenburg, Report 77:13; see also L. E. Larsson, *ASHRAE 1979 Building Thermal Envelope.*

Isberg, J. 1977. "Sandwich Panel Containing Polyurethane Foam, Thermal and Moisture Investigations, Test House in Fiskebaeck," (in Swedish), Chalmers University, Gothenburg, Report 77:14; see also L. E. Larsson, *ASHRAE 1979 Building Thermal Envelope.*

Isberg, J. 1988. "The Thermal Conductivity of Polyurethane Foams," Div. of Building Technology, Chalmers U. of Technology, Gothenburg, Sweden.

Isberg, J. 1989. "Sandwich Panel Containing Polyurethane Foam, Effect of Facing on the Change in Thermal Performance with Time," (in Swedish), Chalmers University, Gothenburg, Report 77:13.

ISO. 1991. DP 10456-1991-07-02 draft proposal "Methods for Determining Declared and Design Thermal Values for Building Materials and Products."

Johannesson, G. and M. Andersson. 1989. Thermal Bridges in Building Structures, Calculation of Thermal Transmission and Surface Temperatures (in Swedish), R34, Swedish Council for Building Research, Stockholm.

Johannesson, G. and O. Aberg. 1981. Thermal Bridges in Sheet Metal Constructions (in Swedish), Lund Inst. of Technology, Report TVBH-3006, Lund, Sweden, pp. 1–76.

Johannesson, G. 1981. Thermal Bridges in Sheet Metal Constructions, Studies in Building Physics, Dedicated to Prof. L. E. Nevander (in English), Lund Inst. of Technology, Report TVBH-3007, Lund, Sweden, pp. 95–114.

Johannesson, G. 1981. Active Heat Capacity, Models and Parameters for the Thermal Performance of Buildings, Lund Inst. of Technology, Report TVBH-1003, Lund, Sweden.

Kabayama, M. 1987. "Long-Term Thermal Resistance Values of Cellular Plastic Insulations," *J. of Thermal Insulation,* 10:286–300.

Karagiozis, A. and M. K. Kumaran. 1993. "Computer Model Calculation on the Performance of Vapor Barriers in Canada Residential Buildings," *ASHRAE Transactions,* 99(2):991–1003.

Karpati, K. K. 1984. "Investigation of Factors Influencing Outdoor Performance of Two-Part Polysulphide Sealants," *J. of Coatings and Technology,* 56(719):57–60.

Karpati, K. K. 1985. "Testing Polysulphide Sealant Deformation on Vises," *Adhesives Age,* 28(5)18–27.

Kashiwagi, D. T. and W. C. Moor. 1986. "Validation of Polyurethane Foam Roof System," *J. of Thermal Insul.,* 10:91–110.

Kashiwagi, D. T. and W. C. Moor. 1993. "The Relationship between Energy Cost and Conservation Measures, Building Design and Insulation Levels," *J. of Thermal Insulation and Bldg. Envelopes,* 16:375–394.

Kashiwagi, D. T., J. P. Nuno and W. C. Moor. 1994. "Optimizing Facility Maintenance Using Fuzzy Logic and the Management of Information," *16th Int. Conf. on Comp. and Ind. Eng.,* March 7-9, Ashikaga, Japan, pp. 404–407.

Kashiwagi, D. T. 1995. "Performance Based Procurement System for Roofing," *J. of Thermal Insul. and Bldg. Envs.,* 19:49–58.

Kashiwagi, D. T. and M. K. Pandey. 1997. "Resistance of Polyurethane Foam Roofs against Hail Impact," *J. of Thermal Insul. and Bldg. Envs.,* Vol. 21, in print.

Khalil, M. A. K. and R. A. Rasmussen. 1986. "The Release of Thichloromethane from Rigid Polyurethane Foams," *Air Pollution Control Association Journal,* 36(2): 190–194.

Khalil, M. A. K. and R. A. Rasmussen. 1989. "The Residence Time of Thichloromethane in Rigid Polyurethane Foams: Variability, Trends and Effects of Ambient Temperature," *Advances in Foam Aging.* D. A. Brandreth, ed., Caissa Editions, Yorklyn, DE.

Knapen, M. and P. Standaert. 1985. "Experimental Research on Thermal Bridges in Different Outer-Wall Systems," Building Physics Lab., Kath. U. of Leuven, presented at CIB-W40 at Holzkirchen.

Kosny, J. and A. O. Desjarlais. 1994. "Influence of Architectural Details on the Overall Thermal Performance of Residential Wall Systems," *J. of Thermal Insul. and Bldg. Envs.,* 8:53–69.

Kosny, J. 1995. "Comparison of Thermal Performance of Wood Stud and Metal Frame Wall Systems," *J. of Thermal Insul. and Bldg. Envs.,* 19:59–71.

Kuehn, T. H. and E. A. B. Maldonado. 1984. "Two-Dimensional Transient Heat Transfer through Composite Wood Frame Walls—Field Measurements and Modeling," *Energy and Buildings,* 6:55–66.

Kumaran, M. K. and M. T. Bomberg. 1985. "A Gamma-Spectrometer for Determination of Density Distribution and Moisture Distribution in Building Materials," *Proceedings of the International Symposium on Moisture and Humidity,* Washington D.C., April 15–18, pp. 485–490.

Kumaran, M. K. 1989. "Experimental Investigation on Simultaneous Heat and Moisture Transport through Thermal Insulation," *Proceedings of Int. Council for Building Research, CIB Congress 1989, Theme II, Volume II,* pp. 275–284.

Kumaran, M. K., M. T. Bomberg, R. G. Marchand, J. A. Creazzo and M. R. Ascough. 1989. "A Method for Evaluating the Effect of Blowing Agent Condensation on Sprayed Polyurethane Foams," *Journal of Thermal Insulation,* 13:123–137.

Kumaran, M. K. and M. T. Bomberg. 1990. "Thermal Performance of Sprayed Polyurethane Foam Insulation with Alternative Blowing Agents," *J. of Thermal Insulation,* 14:43–58.

Kumaran, M. K. 1990. " Experimental Investigation on Simultaneous Heat and Moisture Transport through Thermal Insulation," *Theme II, Volume II, Int. Council for Building Research Studies and Documentation CIB World Congress,* Paris, pp. 275–284.

Lacasse, M. A. 1994. "Advances in Test Methods to Assess the Long-Term Performance of Sealants," *ASTM STP 1254,* Am. Soc. Testing and Mat., Philadelphia, PA, pp. 5–20.

Larsson, L. E. 1978, 1987. "Determination of Thermal Conductivity in Long-Term Tests of Polyurethane Used as Thermal Insulation in Sandwich Panels," Chalmers Tekniska Hogskola, Sweden, May (publication 78:3) and March (publication 87:3).

Larsson, L. E. 1989. "Polyurethane Foam Sandwich Panels, Laboratory Tests on Moisture Migration under Variable Climatic Conditions," *CIB W-40 Meeting,* Victoria, Canada, pp. 1–21 (CTH paper no. 712, publ. 89:3), Jan Isberg, presentation at NRC Canada, May 1992.

Levy M. M. 1966. "Moisture Vapor Transmission and Its Effect on Thermal Efficiency of Foam Plastics," *J. Cellular Plastics,* 2(1):37–45.

Lstiburek, J. W. and J. K. Lischoff. 1984. A New Approach to Affordable Low Energy House Construction, Lincolberg Dev. Corporation and Alberta Department of Housing.

Lstiburek, J. W. 1990. "Insulation Induced Exterior Paint Peeling and Siding Failures," *J. of Thermal Ins.,* 13:205–217.

Lstiburek, J. W. 1992. "Two Studies of Mold and Mildew in Florida Buildings," *J. of Thermal Ins. and Bldg. Envelopes,* 16:66–80.

Lstiburek, J. W. 1994. *Mold, Moisture and Indoor Air Quality, a Guide for Designers, Builders and Building Owners,* Building Science Corporation, pp. 1–224.

Lstiburek, J. W. 1995. "A Case of Cladding Problems Highlights the Need for a Holistic Approach to the Facility Design," *J. of Thermal Ins. and Bldg. Envelopes,* 19: 12–27.

Lstiburek, J. W. and M. T. Bomberg. 1996. "The Performance Linkage Approach to the Environmental Control of Buildings, Part 1: Construction Today," *J. of Thermal Insul. and Bldg. Envs.,* 19:244–278.

Lstiburek, J. W. and M. T. Bomberg. 1996a. "The Performance Linkage Approach to the Environmental Control of Buildings, Part 2: Construction Tomorrow," *J. of Thermal Insul. and Bldg. Envs.*, 19:386–403.

Low, N. M. P. 1990. "The Characterization of Thin Layers of Cellular Plastic Insulations," *J. of Thermal Insulation*, 13:246–261.

Lux, M. E. and W. C. Brown. 1989. "Air Leakage Control," *Proceedings of Building Science Insight 86*, NRCC 29943, January, pp. 13–19.

Maji, A. K., D. Sahpati and S. Donald. 1995. "Experimental Investigation of Tensile Fracture in Polyurethane Foams," *J. of Mat. in Civil Eng.*, 7(4):258–264.

McElroy D. L., R. S. Graves, F. J. Weaver and D. W. Yarbrough. 1990. "The Technical Viability of Alternative Blowing Agents in Polyisocyanurate Roof Insulation, Part 3: Acceleration of Thermal Resistance Aging Using Thin Boards," *Polyurethanes 90*, pp. 247–260.

Mitalas, G. P. 1979. "Effect of Building Mass on Annual Heating Energy Requirements," *2nd Int. CIB Symp. on Energy Conservation in Built Environment*, Copenhagen, 2:7–16.

Mølhave. 1990. "Volatile Organic Compunds, Indoor Air Quality and Health," *Indoor Air '90: Proc. 5th Int. Conf. Indoor Air Quality and Climate*, Toronto, Canada, 5:15–33.

Munroe, J. A., K. Clarke and R. Jung. 1985. "Wood Deterioration in Vegetable Storages Sprayed with Polyurethane Insulation," *Canadian Soc. of Agr. Engineering Meeting*, June 23–27, paper 85-411, pp. 1–19.

NAHB. 1992. *Testing and Adoption of Spray Polyurethane Insulation for Wood Frame Building Construction, Phase 2—Wall Panel Performance Testing*, SPFD/SPI.

Norton, F. J. 1967. "Thermal Conductivity and Life of Polymer Foams," *J. of Cellular Plastics*, 3:23–27.

Norwegian, B.R.I. 1982. Moisture Accumulation in Cellular Plastics, Test Method 135/82, 2 issue.

NRCA. 1996. "Sprayed Polyurethane Foam-Based Roofing Manual," *The NRCA Roofing and Waterproofing Manual*, Fourth Edition, 2:1479–1624.

Ojanen, T. and M. K. Kumaran. 1992. "Air Exfiltration and Moisture Accumulation in Residential Wall Cavities," *Thermal Performance of Exterior Envelopes of Buildings V, Proc. ASHRAE*, pp. 491–500.

Ojanen, T. and M. K. Kumaran. 1996. "Effect of Exfiltration on Hygrothermal Behavior of a Residential Wall Assembly," *J. of Thermal Insulation and Bldg. Envs.*, 19:215–227.

Ondrus, J. 1977. "Thermal Investigations of Sandwich Panel Containing Polyurethane Foam, Test House in Landsbro, and Laboratory Tests," (in Swedish), Chalmers University, Gothenburg, Report 77:12; also L.E. Larsson, *ASHRAE 1979, Building Thermal Envelope*.

Onysko, D. M. and S. K. Jones. 1992. "Air Tightness of Two Walls Sprayed with Polyurethane Foam Insulation," Forintek Canada Corp Report 43-10C-016; or *Journal of Thermal Insulation*, 15:339–354.

Ostrogorsky, A. G. and L. R. Glicksman. 1986. *Aging of Polyurethane Foams, the Influence of Gas Diffusion on Thermal Conductivity*, Oak Ridge National Laboratory, ORNL/SUB/84-9002/2, pp. 1–96.

Pazia, A. P. 1995. "The Dual Challenge of Air Leakage and Thermal Efficiency—The One Step Solution," *Seventh Canadian Masonry Symposium*, McMaster U, Hamilton, pp. 226–237. See also: Chem-Thane SPF manual.

Perrault, J.C. 1978. "Application of Design Principles in Practice," *Seminar/Workshop: Construction Details for Air Tightness*, Nat. Research Council of Canada.

PFCD. 1989. Introduction to Spray Polyurethane Foam Roofing Systems, PFCD-SPF1-4/89.

PFCD. 1990. Spray Polyurethane Foam Systems for New and Remedial Roofing, PFCD-GS1-/90.

Plonski, W. 1965. Thermal Protection of Buildings, Support Information for Designers (in Polish), Warsaw.

Pratt, A. W. 1969. "Thermal Transmittance of Walls Obtained by Measurements on Test Panels in Natural Exposure," *Building Science*, 3:147–169.

Quiruette, R. L. 1985. The Difference between a Vapour Barrier and an Air Barrier, Building Practice Note 54, NRC Canada.

Rand Corporation. 1980. Economic Implications of Regulating Chlorofluorocarbon Emissions from Non-Aerosol Applications, R-2879-EPA.

Report. 1987. Thermal Insulation, Urethane Spray in Place, Technical Subcommittee CGSB 51.23-M, The Society of Plastics Industry of Canada.

Robinson, T. 1992. "Moisture Challenges in Canadian Energy Efficient Housing," *J. of Thermal Insul. and Bldg. Envelopes*, 16:112–120.

RSR. 1986. Performance Criteria for New and Aged Polyurethane Roofs, The Roof Systems Research Committee, sponsored by the predecessor of Spray Polyurethane Foam Division of SPI Inc.

Sandberg, P. I. 1990. "Deterioration of Thermal Insulation Properties of Extruded Polystyrene. Classification and Quality Control System in Sweden," *ASTM STP 1030*, pp. 197–204.

Sasaki, J. R. 1971. Thermal Performance of Steel-Stud Exterior Walls, Building Res. Note 77, NRC Ottawa.

Schuetz, M. A. and L. R. Glicksman. 1983. "A Basic Study of Heat Transfer through Foam Insulation," *Sixth International Polyurethane Conference*, San Diego, CA, pp. 341–347.

Schwartz, N. V. 1989. Value in Use of Thermal Insulation Systems, Internal Report 580, IRC/NRC.

Schwartz, N. V., M. T. Bomberg and M. K. Kumaran. 1989. "Water Vapor Transmission and Moisture Accumulation in Polyurethane and Polyisocyanurate Foams Water Vapor Transmission through Building Materials and Systems: Mechanisms and Measurement," *ASTM STP 1039*, ASTM, pp. 63–72.

Schwartz, N. V., M. T. Bomberg and M. K. Kumaran. 1989. "Measurement of the Rate of Gas Diffusion in Rigid Cellular Plastics," *Journal of Thermal Insulation*, 13:48–61.

Schwartz, N. V. and M. Bomberg. 1991. "Image Analysis and the Characterization of Cellular Plastics," *Journal of Thermal Insulation*, 15:153–171.

Schwartz, N. V. 1991. Presentation to Polyurethane Reseach Commmittee, SPI/NRC.

Seifert, B. 1990. "Regulating Indoor Air," *Indoor Air '90: Proc. 5th Int. Conf. Indoor Air Quality and Climate*, Toronto, Canada, 5, pp. 35–49.

Sherman, M. 1980. "Sampling Faced Foam Insulation Board for Heat Flow Meter Thermal Performance Testing," *ASTM STP 718*, pp. 298–307.

Sherman, M. 1981. "Aged Thermal Resistance (R-Value) of Foil Faced Polyisocyanurate Foam Thermal Insulation Board, Thermal Performance of the Ext. Envelope of Buildings I," *ASHRAE SP 28*, pp. 952–958.

Shu, L. S., A. E. Fiorato and J. W. Howanski. 1979. "Heat Transmission Coefficients of Concrete Block Walls with Core Insulation," *Thermal Perf. of Exterior Building Envelopes*, pp. 421–435.

Silberstein, A and H. Hens. 1996. "Effects of Air and Moisture on Thermal Performance of Insulation in Ventilated Roofs and Walls," *J. of Thermal Insul. and Bldg. Envs.*, 19:367–385.

Sing, S. N., D. Daems and J. P. Lynch. 1995. "Techniques to Assess the Various Factors Affecting the Long-Term Dimensional Stability of Rigid Polyurethane Foam," ICI Reprint WD-507 from *Polyurethane 95 Conference*, pp. 3–22.

Smith, T. L. 1993. "How Did PUF Roofs Perform during the Hurricane Andrew?" *NRCA Professional Roofing*, January, pp. 20–24.

Smith, T. L. 1993a. "Technology Transfer," *NRCA Professional Roofing*, October.

Smits, G. F. and J. A. Thoen. 1990. "Computer Modeling of Heat Transfer through Rigid Polyurethane Foam," *J. of Thermal Insulation*, 14:81–92.

SPI Polyurethane Division k-Factor Task Force. 1988. "Rigid Polyurethane and Polyisocyanurate Foams: An Assessment of their Insulating Properties," *Polyurethane 88*, SPI, pp. 323–337; see also the previous position paper of SPI, Urethane Division Bulletin U 108, SPI, 1979.

SPI (the Society of the Plastics Industry) of Canada. 1992. "Residential Wall Air Barriers Using Cellular Plastic Insulation," *J. of Thermal Ins. and Bldg. Envelopes*, 16: 14–24.

Steinle, H. 1971. "On Behavior of Polyurethane Foams in Refrigerator Cabinets," *Progressive Refrigeration Science Technology, 13th Int. Congr. Refrig.*, Vol. 47.

Swartz, J. 1995. "Better Building Envelopes: Changes to Part 5 of the National Building Code," *Construction Canada*, May, pp. 33–35.

Swedisol. 1981. "Thermal Bridges in Highly Insulated Sheet Metal Constructions," Sweden.

Thorsen, S. H. 1973. "Determination of Expansion Coefficient, Water Intake, Moisture Diffusivity and Thermal Conductivity for Polyurethane Cellular Plastics with Various Moisture Contents," Chalmers Inst. of Technology, Gothenburg, Sweden, Report 73:6.

Tobiasson, W. and J. Ricard. 1979. "Moisture Gain and Its Thermal Consequence for Common Roof Insulations," paper 1361, *Proc. 5th Conf. on Roofing Technology*, NRCA.

Trethowen, H. A. 1995. "Validating the Isothermal Planes Method for R-Value Predictions," *Transactions, ASHRAE*, 101(2): Preprint SD-95-5-1.

Tsederberg, N. V. 1965. *Thermal Conductivity of Gases and Liquids*, The MIT Press.

Tsuchiya, Y. 1988. "Test Method for Rigid Polyurethane Off-Gassing and Its Decay," *National Research Council of Canada, Canadian Polyurethane Conference*, Val-David, Quebec.

Tsuchiya, Y. and J. M. Kanabus-Kaminska. 1993. "Organic Emissions and Their Decay from HCFC-141B Blown Rigid Polyurethane Foam," *World Polyurethane Congress 1993,* Vancouver, Canada, pp. 368–372.

Tye, R. P. 1987. *Assessment of Foam-in-Place Urethane Foam Insulation Used in Buildings,* Oak Ridge National Laboratory, ORNL/Sub/86-56525/1, pp. 1–64.

Valore, R. C. 1980. "Calculation of U-Values of Hollow Concrete Masonry," *Concrete Int.,* pp. 40–63.

Valore, R., A. Tuluca and A. Caputo. 1988. *Assessment of the Thermal and Physical Properties of Masonry Block Products,* ORNL/Sub/86-22020/1, Oak Ridge National Laboratory.

Vanier, D. J. 1993. "Minicode Generation: A Methodology to Extract Generic Building Codes," *CAAD Futures '93: Proceedings of the Fifth International Conference on Computer-Aided Design Futures,* July 7–10, Pittsburgh, PA, pp. 225–239.

Vanier, D. J. 1994. "A Parsimonious Classification System to Extract Project-Specific Building Codes," PhD thesis, Université de Montréal, Montréal, Québec, October.

Wagner, A., L. Preissler, R. Muller and R. Rudolphi. 1984. *Measure or Calculate—Thermal Insulation as the Building Protection Problem* (in German), Bundesanstalt fur Materialprufung (BAM), Berlin, Heft 6, pp. 419–421.

Wiedermann, R. E., N. Adam and R. Kaufung. 1987. "Flame Retarded, Rigid PUR Foams with Low Thermal Conductivity," *Polyurethane World Congress,* pp. 99–103.

Wilkes, K. E and J. L. Rucker. 1983. "Thermal Performance of Residential Attics," *Energy and Buildings,* 5:263–277.

Wilkes, K. E., R. L. Wendt, A. Delmas and P. W. Childs. 1991, "Thermal Performance of One Loose-Fill Fiberglass Attic Insulation," *ASTM STP 1116,* American Soc. for Testing and Materials, pp. 275–291.

Willingham, R. 1991. *Polyurethane Foam Seminar,* Construction Canada, March, pp. 33–34.

Wolf, S., K. R. Solvason and A. G. Wilson. 1966. "Convective Air Flow Effects with Mineral Wool Insulation in Wood-Frame Walls," *ASHRAE,* 72(2):111.

Woods, A. 1988. *In situ Foams as Insulating Air/Vapor Barriers,* Construction Canada, pp. 51–54.

Woods, G. 1987. *The ICI Polyurethane Book,* ICI Polyurethanes, John Wiley & Sons.

Wright, J. R. 1972. "The Performance Approach: History and Status," *NBS Spec. Publication 361,* Vol. 2, *Performance Concept in Buildings; Proc. of Joint RILEM-ASTM-CIB Symp.,* May 2–5, Philadelphia, PA.

Yarbrough, D. W., R. S. Graves and J. E. Christian. 1991. "Thermal Performance of HCFC 22 Blown Extruded Polystyrene Insulation," *ASTM STP 1116,* pp. 214–228.

Zarate, D. A. and R. L. Alumbaugh. 1982. Thermal Conductivity of Weathered Polyurethane Foam Roofing, Naval Civil Engineering Laboratory, Port Hueneme, N-1643, pp. 1–26.

Zhang, J. S., J. M. Kanabus-Kaminska and C. Y. Shaw. 1994. "A Full Scale Test Chamber for Material Emission Studies and Indoor Air Quality Modelling," *ASTM Symp. on Methods of Characterizing Indoor Sources and Sinks,* Sept 24–28, Washington, D.C.

Zehendner, H. 1983. "Performance Criteria and Test Methods for Thermal Insulation Materials Directly Exposed to External Environment," *CIC 1983,* pp. 491–501.

Zehendner, H. 1987. Langzeitverhalten von Polyurethan-hartschaum-Damm-stoffen (Long-term Performance of Polyurethane Foam, in German), Polyure-thane Industrial Association Report 32, June.

Zehendner, H. 1987. "PUR Spray Foam for Roofs—Use in Germany—Standard Requirements—On Site Inspection," (in German), *Polyurethane World Congress 1987,* pp. 282–285.

Zehendner, H. 1988. Report on the Long-term Performance of Rigid Polyurethane Foam in Flat Roofs, Polyurethane Industrial Association Report 34a, October.

NOTE: TECHNICAL INFORMATION AVAILABLE FROM SPFD/SPI INC.

A guide for selection elastomeric protective coatings over SPF, 1994, Stock #A 102

SPF systems for new and remedial roofing, 1994, Stock #A 104

SPF roofing buyer's check list, 1989, Stock #A 105

SPF blisters, their causes, types, prevention and repair, 1989, Stock #A 107

SPF aggregate systems for new and remedial roofing, 1990, Stock #A 110

SPF systems for cold storage facilities operating between −40°F and 50°F, 1990, Stock #A 111

SPF for building envelope insulation and air seal, 1994, Stock #A 112

Moisture vapor transmission, 1994, Stock #A 118

Glossary of terms common to SPF industry, Stock #A 119

SPF estimating reference guide, 1992, Stock #A 121

The renewal of SPF and coating roof systems, 1994, Stock #A 122

Thermal barriers for SPF foam industry, 1995, Stock #A 126

Maintenance manual for SPF roof systems, 1995, Stock #A 127